A SHEARWATER BOOK

Seeing Things Whole

Other titles from William deBuys

Enchantment and Exploitation: The Life and Hard Times of a New Mexico Mountain Range (1985)

River of Traps (1990)

Salt Dreams: Land and Water in Low-Down California (1999)

Seeing Things Whole: The Essential John Wesley Powell is published in cooperation with the William P. Clements Center for Southwest Studies, Southern Methodist University.

Seeing Things Whole

◆

The Essential

John Wesley Powell

◆

Edited by William deBuys

Island Press / SHEARWATER BOOKS

Washington • Covelo • London

ISLAND PRESS is a trademark of The Center for Resource Economics.

Library of Congress Cataloging-in-Publication Data
Seeing things whole : the essential John Wesley Powell / William deBuys
p. cm.
ISBN 1-55963-872-9. — ISBN 1-55963-873-7
1. Powell, John Wesley, 1834–1902. 2. Colorado River (Colo.–Mexico)—Description
and travel. 3. West (U.S.)—Description and travel. 4. Grand Canyon (Ariz.)—
Description and travel. 5. West (U.S.)—Environmental conditions. 6. Arid regions—
West (U.S.) 7. Powell, John Wesley, 1834–1902—Philosophy. 8. Explorers—United
States—Biography. 9. Geologists—United States—Biography. 10. Conservationists—
United States—Biography. I. Title.
F788 .D34 2001
550' .92—dc21
2001001809
British Cataloguing-in-Publication Data available.

Printed on recycled, acid-free paper
Manufactured in the United States of America
10 9 8 7 6 5 4 3 2 1

Contents

Part VI *Advice for the Century*

Part VII *A Philosopher for Humankind*

MAPS AND PHOTOGRAPHS

———◇———

Maps

1. Map of the Grand Cañon of the Colorado, Showing Route Traveled by Major Powell
2. Map of the Forest Lands of the Arid Region
3. Map of Utah Territory Representing the Extent of the Irrigable, Timber, and Pasture Lands (1878), entire
4. Map of Utah Territory Representing the Extent of the Irrigable, Timber, and Pasture Lands (1878), detail
5. Linguistic Stocks of American Indians North of Mexico (1891), entire
6. Linguistic Stocks of American Indians North of Mexico (1891), detail
7. Arid Region of the United States Showing Drainage Districts (1891), entire
8. Arid Region of the United States Showing Drainage Districts (1891), detail

Photographs

1. Major Powell Inquires for the Water Pocket
2. Major Powell and Jacob Hamblin in a Paiute Council on the Kaibab Plateau
3. Major Powell and Companions Sharing a Meal on the Trail
4. Major John Wesley Powell in His Office in the Adams Building
5. Farewell Dinner for Powell, 1894

The idea for this book grew from a morning's confusion. At my farm several summers ago I was rereading two favorites, *Heart of Darkness* and *Beyond the Hundredth Meridian.* Grabbing a cup of coffee between irrigation chores, I settled into a chair where the slanting sunlight was strong and warm. I picked up a book, opened to the mark, and began to read, expecting to join Marlow on his voyage up the Congo. Instead I met John Wesley Powell headed the other way on the Colorado.

My copy of *Heart of Darkness* was contained in an anthology called the *The Portable Conrad.* Having suddenly encountered Powell between what I thought were Conrad's covers, I readily imagined a similar anthology that would collect the most important work of the redoubtable one-armed Major. As a student of western American history and a participant in occasional environmental wrangles, I regularly consulted Powell's *Report on the Lands of the Arid Region.* Even more often I dug from my files photocopies of the three articles Powell published in *Century Magazine* in 1890, which offer the fullest development of his vision for the West. Powell's work was more than a century old, but in all that time the themes and intensity of arguments about the allocation and use of western lands had scarcely changed. In all that time, no one had emerged who understood the underlying issues of the West better than Powell, and no one, it seemed to me, had offered ideas for addressing them that were as original or as profound as his. The more I thought about it, the more it seemed that the contemporary West needed easier access to the wisdom of John Wesley Powell.

Walt Coward, a marvelous gentleman and serious student of the Southwest, then working for the Ford Foundation, agreed that an anthology of Powell's writing might have merit. He graciously invited a proposal. Ed Marston and Hal Rothman spoke in support of it. A small grant came my way, enough to render my commitment to the project irrevocable. Charles

Wilkinson, Donald Worster, and Cherie Scheick advised me on selections. Charles's former research assistant, Scott K. Miller, helped chase down obscure material from the Bureau of Ethnology. John Herron and later Will Barnes located other materials, and Will, to whom I owe special thanks, proofed and read with critical care the digital files that Leslie North, using all kinds of electronic magic, conjured from the imperfect photocopies we took to her. In the early stages of the project, my long-time friend David Williams helped clear up copyright questions.

Even with the help of so many good people, a large amount of research and writing remained. At an opportune moment, the Clements Center for Southwest Studies at Southern Methodist University (SMU) offered a fellowship that allowed me to work on the project almost without interruption. Moreover, once I arrived at SMU, David Weber, the center's benevolent czar, and Jane Elder, who kept the operation running, were unstinting in their kindness to their visitor from New Mexico and did much to make my year at SMU productive and pleasurable. I cannot thank them enough. Sherry Smith, another mainstay of the Clements Center, and Bob Righter, her not-quite-retired historian husband, also helped make SMU the right place to complete a large project. Across a corner of the main quad from the Clements Center, David Farmer and Kay Bost applied the resources of the fabulous DeGolyer Library to providing several of the maps and illustrations in this volume, and, a short distance from campus in another direction, Delbert and Waunell Hughes generously provided me a place to stay on my weekly trips to Dallas. My warmest thanks to all.

Sometimes research takes strange and wonderful forms, and early on I became convinced that to write about Powell I needed to do more than merely gaze on the Colorado River and Grand Canyon. Don Usner and Deb Harris gave me the chance to immerse myself, sometimes literally, in one of the planet's most spectacular places. After 18 days rafting 225 river miles from Lees Ferry to Diamond Creek, I had a new appreciation for the river, the canyon, and Powell. My thanks to Don, who also reviewed parts of the manuscript, to Deb, and to all my companions on that most wonderful trip.

I am also especially grateful to Don Worster, who kindly reviewed the penultimate draft and caught many errors. Those that remain are, of course, mine alone. Dan Flores helped with periodic advice and with the loan of a copy negative of Powell's marvelous map of western watersheds. Other illustrations came courtesy of the National Anthropological Archives at the Smithsonian Institution, where Paula Fleming was especially helpful. Stacy D. Allen of the National Park Service provided advice and information on Powell's experience in the Battle of Shiloh. Back in Santa Fe, Bill Miller patiently guided me through questions about measuring and administering water rights, and Adriana de Julio and Katie deBuys helped with eleventh-hour transcriptions and other office work. Even later than the eleventh hour, Janice St. Marie provided vital advice on cover design. I thank them all.

Turning a heap of pages into a book takes a good editor, and I have been fortunate to work with one of the best. Jonathan Cobb of Shearwater Press has been relentlessly reasonable, insightful, gentle in his criticisms, and patient.

On a personal note, since Powell's commitment to exploration and service marks an impressive pair of beacons for an adult life, I would like to dedicate my efforts in this volume to my soon-to-be adult children, Catherine and David.

Finally, though, drawing attention to the labors of a compiler and annotator may be inappropriate. This is not my book. It belongs to John Wesley Powell. It seems a puny gesture to express thanks to the memory of a giant such as he, but then again, everyone who lives in the West or wrestles with issues of land use and governance owes Powell a debt of gratitude. May this book bring a new generation of readers to his work and may it also inform the efforts of those who grapple with the issues he did so much to help us understand.

—◇—

Seeing Things Whole

John Wesley Powell was an American original. He was the last of the nation's great continental explorers and the first of a new breed of public servant: part scientist, part social reformer, part institution builder. Although his insights into the lands of the American West remain unsurpassed, the trap in thinking about him has been to consider his work merely regional in interest and influence. Certainly the focus of his mature life was the West. No one before or since has striven so assiduously to comprehend the complexities of that vast land, and none has come away with such depth of understanding. But Powell was more than a western curiosity, more than a merely eye-catching historical sunset. What his work and life reveal to us is a way of thinking about land, water, and society as parts of an interconnected whole.

Virtually alone among his late-nineteenth-century contemporaries, he saw that the character of western lands would shape—and in turn be shaped by—the way in which those lands were settled. He further saw that the result of that interaction would ramify onward for generations and would have profound consequences for the land and for American society. Powell was America's first great bioregional thinker, and the lessons he taught are lessons we still are at pains to learn.

Powell emphasized two things that many of his contemporaries, especially westerners, found troubling. The first was that the lands of the West were not an empty stage that westering Americans could people and build on as they wished. The land had limits, Powell said again and again, and one of them was its aridity, which the settlers of the region would ignore at their peril. Many westerners and would-be westerners, from homesteaders to senators, did not like that kind of talk. It sounded too negative for what seemed to be a boundless American future, a future in which the West, as everybody knew, would play a central role. Still, they listened to Powell, and they almost adopted what he advocated because he was clear, he knew the facts, and he had the authority of a proven national leader. In the Civil War and later in his explorations of the Colorado River, Powell had stared death in the face, fed it his right arm, and then defied it again by riding a 900-mile cataract through territory known to the rest of the world only by rumor, legend, and tiny scraps of fact.

In 1869, when Powell embarked with nine men and four boats on his exploration of the Colorado River, the little that was known about the downstream country only magnified the great deal that was conjectured: The canyons of the Colorado River country were difficult and perilous. To enter was to risk all. Powell did not hesitate a moment. He leapt in, beat the odds by a hair, and not only came out alive but returned to society with hard-won wisdom. Although celebrated for his achievement, he became more than a celebrity. His adventure followed the arc of an American Jason or Odysseus. By the standards of both myth and history, he was a genuine hero.

The second troubling thing Powell kept saying was that the way in which people settled the West would have irremediable consequences. Provide the wrong institutions, the wrong systems for survey and land tenure, the wrong basis in law for holding water rights, and the results would be suffering, betrayed ideals, loss of wealth, and the erosion of democracy. Powell was right about this, too, and probably even his enemies at some level knew he was right.

The alternative Powell proposed involved extensive surveys to classify lands, new laws to govern their use, and new structures of local government to nurture the growth of communities in balance with the capacity of the land. At the heart of his vision lay an appreciation that the West was too dry to support the kind of agriculture that had provided a foundation for settlement of the East. A 160-acre farm in Kentucky, Illinois, or even eastern

Kansas might grow enough corn and pork to support a family, thanks to good soil and ample rainfall. But westward the climate grew drier, and 15 inches of rain in eastern Colorado, 12 in Idaho, or 8 in Arizona could hardly grow what 35 did back east. In those western regions the notion of a 160-acre farm, which the Homestead Act of 1862 enshrined as a national ideal, became not merely laughable but cruel, deceptive, and malignant. As Powell repeated countless times, the Homestead Act's promise of free land for family farms became a hoax in the West. In what he called the Arid Lands, the act guaranteed suffering and sorrow by encouraging people to stake their all on a gamble they were sure to lose. Less than 20 reliable inches of rainfall would not produce a marketable crop of corn or any other crop, no matter what the acreage. To expect otherwise was to embrace ruin. And yet, with the government's encouragement, farmer after farmer and family after family went west and tried to eke out a living from hardscrabble, arid home-steads. Legions of settlers failed in the attempt, broken in wealth and spirit, and their lands fell to the control of speculators and corporations.

Powell argued that there was an alternative to this national tragedy of injury and betrayal. It began with irrigation, which allowed agriculture to flourish on even the driest land. But for irrigation to be developed equitably and to flourish, water resources had to be quantified and irrigable lands susceptible of irrigation identified, none of which could be accomplished while settlement proceeded helter-skelter. Moreover, irrigation agriculture required that farmers cooperate closely in operating and maintaining their ditches, reservoirs, and other infrastructure. These interactions would necessitate new institutions.

The rest of the western landscape also needed a fresh approach. Vast areas where irrigation was impractical might support livestock, but the 160-acre homestead was pitifully inadequate as a ranching unit. Similar prob-lems attended the allocation and use of forested lands, which were doubly important because they were the main source of the water that might be used in irrigation. Powell saw that each of these different classes of land should be treated individually and that the treatment of each must comple-ment that of the others. Only then might a prosperous, stable, and just society be assured. And all this, he understood, would require a fundamentally new approach to land tenure.

Powell's call for a carefully planned approach to settlement was logical and well reasoned, but in the view of his opponents it was bad politics. Powell, they said, wanted legions of farmers, ranchers, traders, and speculators, individuals and families—almost everyone involved in westward migration and settlement—to stop in their tracks, retrace their steps, and come around again by a different route. If their arguments became tinged with hysteria, they nevertheless had a point, which Powell did not dispute. Implementing his plan would have disrupted the frenetic land-trading, farm-making, public-domain–settling economy of the entire West (and for a few months in 1889 and 1890 it did just that). Powell knew that the medicine to cure society's ill relationship to the arid lands would have a bitter taste, but exactly how much disorder and delay he was willing to accept is open to debate. His ideas were always a work in progress, and he frequently acknowledged that details of his plans remained to be worked out. But his themes never varied. Again and again he urged westerners to adapt to the land, to organize institutions that would cultivate democracy at local and regional scales, and to reform the laws that undermined the health of the land and society. And he urged always that these actions not be piecemeal, but that they be unified in a new and integrated approach to the settlement of the arid West.

People in the West and in the nation pondered Powell's proposals, but ultimately they rejected his ideas, not because they considered his analysis wrong—that was never the issue. The problem was that the path Powell urged would take too long and entail too much change. People would have to reorder the way land was distributed, they would have to form new and (for the time) strangely communal institutions for community cooperation, and they would have to wait while experts quantified water resources and allocated land to different uses. But westerners did not want to wait, and they did not want a revolution in land tenure. They wanted to get on with making homes and money by the means most familiar to them. They rejected Powell's alternatives, saying in essence, "We'll keep on going the way we are going, and if we get stuck, we'll get stuck, and if we get somewhere, there we'll be."

Where they ended up is the quite imperfect present, where westerners and all Americans and their global neighbors are today. The West of the American present is a turbulent, less-than-ideal place not simply because its people did not take Powell's advice but because of a thousand and one occurrences that intervened and in turn spawned myriad effects. And yet the West of the present would be a better place if the West of the past had earlier and more completely followed Powell's ideas. The proof of this proposition is all around us: As a society we have traveled a fair distance, falteringly and gracelessly perhaps, in the direction Powell bade us go more than a century ago. Our journey has been incremental, goaded by necessity and crisis. We might think of it as a long course of painful recuperation, and we might wonder how much agony might have been prevented had we taken our medicine sooner, as he advised.

With each generation the grain of history grows finer, and we see greater complexity in the origin of things. The existence in the contemporary West of irrigation districts on nearly every river, of public timber and grazing commons (albeit in a more federalized form than Powell would have wished), of systems of leases for use of those lands, and (at least for the moment) of increasing community influence in managing those commons—we cannot say that these phenomena grew directly from Powell's recommendations. Nor can we easily attribute to Powell the recent emergence of regional, often watershed-based interest groups and management approaches that take shelter under the terms "ecosystem management" or "bioregionalism." But the fact remains that these realities, however imperfect they may be and however long it may have taken us to get to them, tend in their slow way toward the vision for western settlement that Powell propounded in the last quarter of the nineteenth century. The conclusion to be reached is not that events followed Powell's plan but that his integrated understanding of the requirements of western lands was profound, if not prophetic, and that we can still learn from him. Which leads us to a final question: If much that he recommended has come to pass, should we not closely reevaluate the elements of his vision that remain unimplemented? It is because of the endurance of his vision that Powell is not just a regional curiosity. He stands as a model of holistic thinking, appropriate to any land or era.

But Powell did more than think and propound. Having striven to grasp the entirety of the mosaic of his time and place and having captured it with uncommon accuracy, he acted on the knowledge thus earned and never shied from the conflicts and difficulties to which it led. This is another reason for knowing and studying Powell. Few leaders of any time have better combined the life of the mind with a life of action.

Powell was a man of contradictions. He can fairly be called a pragmatic idealist, an elitist democrat, and an antifederal federal bureaucrat. This causes consternation in certain quarters. To suggest, as do some of his critics, that to be so riven, to be more complicated than a merely pleasant dinner companion, constitutes a character fault is to trivialize the actual, gnarly difficulty of contending in real terms with real issues. Powell lived on the playing field of American public life, not safely in the bleachers. His ideas were challenged and debated at every turn. Only the most rigid ideologue or a simpleton can happily proffer simplistic, mechanically consistent answers to questions as intricate and sweeping as those that challenged Powell. His true measure is to be found not in the final tidiness of his ideas but in the degree to which he grappled with the contradictions and ambiguities of the actual world around him and then wrestled those tensions into an approximation of harmony.

To appreciate Powell we must keep our frame of reference broad. He was a self-taught polymath who began as a naturalist and explorer and became a dozen different things: surveyor, geographer, geologist, linguist, ethnographer, anthropologist, philosopher, reformer, and institution builder. He achieved all this thanks to qualities of persistence and candor that were as unadorned as his humble origins. As his biographer Wallace Stegner put it, "His homemade education fitted him to grasp the obvious and state it without embarrassment—he had not been educated into scholarly caution and that squidlike tendency to retreat, squirting ink, which sophisticated learning often displays."[1]

How Powell came by his astonishing self-confidence is impossible to say, but there is no question about the homemade quality of his education. He was born in Mount Morris, a village in western New York near the Genesee River, in 1834. His parents, Joseph and Mary Dean Powell, had arrived in the United States from England only 4 years earlier, and, like many Americans of their time and ours, they seldom stayed in one place for long. Joseph Powell, a tailor, farmer, and Methodist exhorter, moved the family to southeastern Ohio in 1838, to southern Wisconsin in 1846, and thence a few miles across the state line into northern Illinois in 1851. At each stop, young John Wesley received the irregular instruction of frontier schools and the home schooling of his parents, which included abundant exposure to the systematic teaching—the method-ism—of the John Wesley for whom he was named. On occasion Powell also benefited from the guidance of that rarity in frontier communities, the educated neighbor, but as he grew older he came increasingly to depend on himself as his own best teacher, especially in natural history, which exerted on him an uncommonly strong pull.

Family needs and farm work drew him away from school before he was a teenager, but periodically he escaped indenture to plow and axe and attended classes at several of Illinois's fledgling colleges. He also indulged his wanderlust with expeditions to collect plants, animals, and mineral specimens—all manner of natural history curiosities. These forays took him ever farther from home, and he rambled far and wide through the woods and prairies of the Middle West, much as another famous preacher's son, John Muir, would soon do. Powell also took to the region's rivers and rowed or floated many a long stretch, even descending the Mississippi in 1856 all the way to New Orleans, enacting an American pilgrimage much honored in fiction and fact. Entering manhood, Powell faced two options for making his way in the world: farming and teaching. With his prodigiously restless mind, his choice was simple. By 1858 he was keeping school

---

[1]Stegner, *Beyond the Hundredth Meridian,* 152.

in Hennipen, Illinois, and had been elected secretary of the Illinois State Natural History Society. He was also paying increasing attention to a young woman 2 years his junior who lived in Detroit. She was his cousin, Emma Dean, the daughter of his mother's half-brother.

When the outbreak of the Civil War convulsed the nation, John Wesley Powell, an ardent abolitionist, responded to the first call-up of volunteers. He enlisted as a private, was quickly elected sergeant by virtue of his education, and soon afterward received a lieutenant's commission to fill a vacancy. He straightaway traveled to Chicago to obtain an officer's uniform, and he made additional use of the trip by buying books on building fortifications and other aspects of military engineering. With characteristic resolve and independence, Powell undertook to make himself expert in a new field of knowledge.

He put his knowledge to use in designing and building defenses for the Union stronghold at Cape Girardeau, Missouri, and when, with construction partly complete, Brigadier General Ulysses S. Grant came to inspect the Union encampment, Powell entered into one of the most important relationships of his life. Even as his friendship with Grant began, the general was immediately instrumental in allowing Powell to advance an even more important relationship. He gave Powell leave to travel to Detroit, where the 27-year-old lieutenant arrived on November 28, 1861. That evening, John Wesley Powell and Emma Dean were married.

When Powell returned to Cape Girardeau, Grant commissioned him a captain of artillery, and it was in that capacity, commanding Battery F, Second Illinois Light Artillery, that Powell embarked by steamboat the next March on a campaign up the Tennessee River. As sometimes occurred early in that conflict, when war was less than total and armies fought only armies, Emma followed not far behind, also traveling by boat. In the first days of April, Powell's unit joined a large force assembling beside the river at a place called Pittsburg Landing, near Shiloh Church. Although the army had gathered there for a considerable time, Powell noted with concern that it had failed to dig in or erect defenses.

At dawn on April 6, elements of a Confederate force numbering 43,699 attacked the ill-prepared and slightly smaller Union army of 39,830.[2] Five hours of savage fighting forced the two most forward Union divisions to fall back with heavy casualties. The retreat threatened to become a rout until Yankee resistance stiffened and Confederate momentum stalled in heavy, brushy terrain near a sunken road in the center of the battle, a place the Confederates called the Hornet's Nest. It was there that Powell had positioned his battery.

Powell had begun the day well to the rear, awaiting orders and listening to the distant roar of battle. Battery F was a new unit in the general encampment, unassigned to any division, and no assignment or specific orders came now. By mid-morning Powell could stand inaction no longer. Under his own initiative, he moved his unit forward to render assistance on the Union right but was soon forced back with the loss of one of the six light cannon he commanded. With his five remaining 6-pounders he reported to General William H. L. Wallace, who directed resistance at the Hornet's Nest, in the Union center. Aware that Grant had ordered his troops to hold the position at all costs, Powell unlimbered his guns in a dense oak thicket.

Along with other batteries similarly positioned, Powell's battery kept up steady fire through the afternoon, changing position several times. Together with nearly 6,000 Federal infantry, they repulsed repeated Confederate attempts to dislodge them. By 4 o'clock, however, their position had become desperate. The Confederates not only beset them from the front but also were rolling back the Union flank, threatening to envelop the Hornet's Nest from the left. Wallace directed Powell to redeploy two guns to buttress resistance on the crumbling left, and Powell moved his section to engage a Confederate battle line across an open field. Many years later, Powell recalled,

---

[2] It is difficult to find detailed agreement among the many descriptions of the Battle of Shiloh. This brief account is drawn from information generously provided by Stacy D. Allen, park historian at Shiloh National Military Park (letter of December 10, 1999).

> Soon I discovered that there was a line of men concealing themselves in the fence and I dismounted and pointed one of the pieces along the fence loaded with solid shell. As I raised my hand for a signal to the gunners to stand clear of the recoil[,] a musket ball struck my arm above the wrist which I scarcely noticed until I attempted to mount my horse.[3]

The minié ball burrowed through Powell's forearm, shattering the bones. General Wallace soon appeared and "picked me up, for he was a tall athletic man, and put me on my horse and directed the sergeant to take me back to the landing." Powell and the sergeant rode back along a slender corridor of Union-controlled ground, receiving fire from both sides: "As we rode certainly hundreds, perhaps thousands of shots were fired at us, none of which took effect." Powell was "very pale" when he reached the landing, and friendly hands helped him from the horse and placed him on a boat that carried him downriver to the hospital tents at Savannah, Tennessee. It was there that Emma found him.

Resistance at the Hornet's Nest collapsed not long after Powell's escape, but the stubborn fighting there, abetted by rough topography, physical exhaustion, and the rallying of Union reserves, helped delay the Confederates long enough for Federal troops to secure their lines for the night. At dawn the next morning a reinforced Union army successfully counterattacked, and the Confederates withdrew. The Battle of Shiloh produced a grim harvest of carnage, leaving nearly 24,000 killed, wounded, or missing. If either side benefited, it was the Union, for the toll of attrition, being equally shared, was more damaging to the less populous South. Moreover, Grant's army still held a position on the Tennessee River from which it threatened rail links vital to the Confederacy.

Two days after the battle Dr. William Medcalfe, a druggist in civilian life, inexpertly amputated Powell's right arm. Powell was fortunate that Emma was close at hand to nurse him through fever and trauma back to health.

---

[3] J. W. Powell to Col. Cornelius Cadle, Chairman of the Shiloh National Military Park Battlefield Commission, May 15, 1896. Courtesy of Stacy D. Allen.

The stump of his arm, however, gave constant pain, and would soon require additional surgery.

After a convalescence of several months and a recruiting tour in the North, Powell returned to active duty, fighting on through the Vicksburg campaign in 1863 and at Nashville the next year. He left the army in January 1865 after the outcome of the war was no longer in doubt.

It is worth considering what effect the Civil War had on Powell and others of his generation, north and south. When the shooting stopped and soldiers returned home, they tried to resume the lives they had led before. But for many, the effort was in vain. Before long, Powell and countless others looked west toward a land that seemed to embody challenge, freedom, and opportunity.

The destiny that a previous generation had called manifest was for Powell and his contemporaries something even surer, not just manifest but assured, and yet it remained largely unrealized. The work of subjugating native peoples—for that is how they saw it, and few of Powell's countrymen held any illusions about the task—was incomplete, and the work of settlement and nation building in the western territories had barely begun. Men like Powell, who had fought and suffered over the question of whether the United States should remain united, came out of the Civil War with a sense of national purpose virtually welded into their bones.

Strong as it was for his generation, the sense of national mission was particularly strong for Powell, and we may not be far off the mark to think that it attached in his psyche near the place where the missionary fervor of his father had failed to adhere. But this was not his only inheritance from the war. On a practical level, the army had taught him a great deal about large organizations and the bureaucracy it took to run them, and he had learned even more about directing large, logistically complex operations. On another level, he had learned much about himself. He had survived a grievous wound, faced the charges and bombardment of enemy troops, built bridges across swamps in the dark of night, and hauled cannon through slit trenches to impossible positions. Nothing he experienced during the war, not even the loss of his arm, diminished the self-confidence with which he had entered it. Indeed, as rest and peace restored his strength in 1865, John Wesley Powell felt ready for any challenge. Soon he would tackle the biggest one that the great western wilderness could offer.

But first there were practical matters. Powell secured a professorship at Illinois Wesleyan University in Bloomington and soon also began teaching at nearby Illinois State Normal University. His courses covered the full spectrum of the natural sciences, from geology to biology, and before long he began lecturing to public audiences as well. He had never ceased his studies during the war. Even among the trenches of Vicksburg, he collected the fossil mollusks his men unearthed, and he made notes on the erosion of the barren no-man's land between the armies. Back in Illinois he was free to continue his investigations without hindrance, if only funds might be found. Powell pressed ahead with the energy of an entrepreneur. An offer to make collections of Rocky Mountain flora, fauna, and minerals for the young Illinois Museum of Natural History, as well for the two universities he served, led to lobbying, and the lobbying led to a grant of $500 from the state legislature and appointment as the museum's curator. He visited General Grant in Washington and came away with permission to draw rations at western army posts. Various railroads and stagelines donated transport for his party and their baggage.

And so away Powell went in the summer of 1867 to the mountains of Colorado with a motley crew of amateur naturalists and undergraduates. One of the highlights of the expedition was the party's ascent of Pike's Peak, in which Emma Powell participated, clad in multiple petticoats and a dress that reached to her boot tops. She may not have been the first white woman to make that climb, as she and others speculated, but her rugged spirit placed her in rare company. The next year Powell led a second expedition to Colorado with a greater number of students, this time to the region of present-day Rocky Mountain National Park. There he and several others made the first ascent of Long's Peak, which towers over Estes Park. At the close of the summer season, Powell again shipped back to Illinois assorted boxes and crates of taxidermied animals, pressed plants, and rock samples. He also sent back students whose worlds had been enlarged. The two Colorado trips had initiated something new in American higher education, for the tradition of summer field study begins with Powell.

Powell's ambitions no longer lay with teaching, however. Rather than return to Illinois, he, Emma, and a few others established a camp on the White River, west of the Continental Divide, and Powell spent the winter exploring the canyons of the White, Green, and Yampa rivers, all of which gave their waters to the mighty and as yet little-known Colorado system. Powell then began to plan in earnest a descent of the river that he hoped to make the next summer. There was no time to lose. The army, which had long played a leading role in scouting the West, was sure to launch an exploration of the unknown country soon, and Powell hungered to be the first to tap the riches of discovery that it was certain to yield. He later wrote, "The thought grew into my mind that the canyons of this region would be a book of revelations in the rock-leaved Bible of geology. The thought fructified, and I determined to read the book."[4]

Powell's famous 900-mile descent of the Colorado River, from May through August 1869, is one of the epic adventures of American history and provides the starting point for this collection of Powell's work. Nothing in Powell's life held more consequence. The exploration of the Colorado was the main source of Powell's future fame and of much of his later influence in the world of policy and government. It was also a defining, vital episode in the development of his thinking about the arid lands of the West. This anthology of Powell's writing, in Part I, begins with a pair of letters he dispatched a little more than a month into the expedition from the Uinta Indian Agency, a remote outpost in the wilds of northeastern Utah. The letters are part of the sole progress report Powell was able to send back about his exploration, and, reprinted by the *Chicago Tribune, New York Times,* and other widely read periodicals, they greatly whetted the public appetite for future news of Powell's adventure. In them we learn of the explorers' severest early trial and gain insight to the strains that eventually fracture the expedition.

---

[4] As quoted by Rabbitt, "John Wesley Powell," 3.

In Part II we rejoin Powell and his men at the mouth of the Grand Canyon and accompany them down a raging river through the most colossal fissure in the crust of the earth. The selection is drawn from the 1895 edition of *Canyons of the Colorado,* which in turn was derived from the extended report of the exploration that Powell submitted to Congress in 1875. It includes Powell's description of the expedition's hair-raising passage through the rapids of the inner canyon and his account of its breakup when three of its members, refusing to risk their lives further in the terrifying whitewater, elected to leave the boats and set out for civilization on foot across the wilds of southern Utah. Some time after his own safe return, Powell learned that they died in the attempt.

Part III is drawn from Powell's accounts of his subsequent exploration by horseback of the Colorado Plateau, including his efforts to learn the fate of the three who died after leaving him at the river. In these selections we are treated to some of Powell's best ethnographic writing. In his descriptions of Mormons, Paiutes, and Hopis we get a sense of his unflagging energy and curiosity, the acuity of his powers of observation, and his profound interest, as a nascent anthropologist, in the wealth and diversity of human cultural experience.

Part IV consists of key chapters from one of the most extraordinary documents in American history. Powell's *Report on the Lands of the Arid Region* is an analysis, completely without precedent, of the character of the arid lands of the West and of the legislative and legal reforms necessary for proper settlement of the region. It presents a blueprint for a democratic and egalitarian West and for a society grounded in a realistic appraisal of the environment on which it depends. No bolder or more original treatment of the West—or of any other region—has ever existed.

Report on the Lands of the Arid Region helped moved Powell to the forefront of national debates on the use of the public domain and on the role of the federal government in developing western states and territories. In Part V we encounter three of his many important contributions to those discussions. In one of them he battles popular misconceptions about the potential for "gentling" the harsh climate of western lands; in another he confronts widespread public anxiety about dam building in the aftermath of the disastrous Johnstown Flood; and in the third he advises delegates to the Montana constitutional convention how best to protect and use the water resources of their soon-to-be state. More importantly, in each case he goes beyond his immediate purposes to explore the challenge of conforming settlement to the imperatives of the land.

Powell's ideas about the arid lands reach their fullest development in Part VI. The three articles presented there first appeared in *Century Magazine* in March, April, and May 1890, and they have not been reprinted until now. Though less well known than the *Report on the Lands of the Arid Region,* the *Century* articles build on that earlier effort and constitute a further development of Powell's views on the arid lands, their settlement, and the allocation and use of their resources. In them (and not in the *Report on the Lands of the Arid Region*) he elaborates the most revolutionary of all his recommendations, arguing that the arid lands should be organized into watershed commonwealths, with each commonwealth governed by resident citizens whose interlocking interests create the checks and balances essential to wise stewardship of the watershed. This was the essence of Powell's final bioregional vision for the West. Had it been proposed in a less compulsively individualistic age, something like it might well have been adopted, and with its adoption, the West of today would be a far different place, less dominated by grand hydraulic works, less oligarchic politically and economically, less battered ecologically, perhaps less wealthy in the short term and less grand in industrial accomplishment, but more durable in the cohesion of its polity and in the marriage of its communities with the land.[5] This, at least, would be the hope, and in the long span of the future there may still be opportunities to move farther in the direction Powell charted more than a century ago.

The anthology's final selections, in Part VII, help to present Powell in something approaching the full diversity of his interests and achievements.

---

[5] Donald Worster has elaborated on these possibilities on a number of occasions. See his *Rivers of Empire,* 329–35; "The Legacy of John Wesley Powell," in Worster's *An Unsettled Country* and also collected in Rothman, *Reopening the American West;* and most importantly Worster's biography of Powell, *A River Running West.*

In these excerpts from his formal anthropological writings, we glimpse the deepest foundations of his thinking about the nature of humankind and of his general philosophy of life. We also see him fully engaged in one of the most heated intellectual debates of the final decades of the nineteenth century, a defining period for the American West and for all of America. This was the controversy aroused by Herbert Spencer's doctrine of extreme economic and political individualism, which became immensely popular in the United States and was derided by critics as "social Darwinism." The selections in Part VII include Powell's repudiation of that doctrine and his contrasting expression of "faith in my fellow-man, towering faith in human endeavor, boundless faith in the genius for invention among mankind, and illimitable faith in the love of justice that forever wells up in the human heart."[6]

Powell's faith in humankind never lies far from the surface in either his writings or his actions. The same might be said of several other themes that run through his endeavors. By now it should be clear that Powell's hunger for new knowledge was inexhaustible and that he never hesitated to plunge into new areas of investigation and learning. He taught himself what he needed to know to become a surveyor and topographical engineer, which is to say a mapmaker who gathers his own data by measuring the land. He also taught himself to become a geographer and geologist, and he did so to such a degree that he virtually invented the subdiscipline of geomorphology, the study of the evolution of landforms.

Among Powell's most original and enduring formulations is the concept of the "base level of erosion"—the idea of a physical limit to the downcutting of rivers and creeks. He also divined the processes by which valleys are formed, and he classified them accordingly. A "consequent" valley, in his terminology, derived from the existing "corrugation" of the landscape, which determined the course of the river that formed the valley. A "super-

---

[6] See page 358.

imposed" valley took its shape from geological strata that were present when the valley was young but that had since eroded away, leaving the form of the valley to persist. And an "antecedent" valley formed from a river that had established its course prior to more recent geologic uplift. The canyon of the Green River through the Uinta Mountains of northeastern Utah was famously described by Powell as an antecedent valley:

> We may say, then, that the river did not cut its way *down* through the mountains, from a height of many thousands of feet above its present site, but, having an elevation differing but little, perhaps, from what it now has, as the fold was lifted, it cleared away the obstruction by cutting a cañon, and the walls were thus elevated on either side. The river preserved its level, but mountains were lifted up; as the saw revolves on a fixed pivot, while the log through which it cuts is moved along. The river was the saw which cut the mountain in two.[7]

This is clear, uncluttered writing, and it perfectly reflects Powell's clear and uncluttered way of seeing large landscapes. Ignoring superfluities, he cut toward the core of the issue at hand, paring away inessential data until he grasped the matter in its most fundamental terms. Having once grasped a truth, he never let it go, as his adversaries repeatedly and grudgingly admitted. When Powell the advocate and author of the arid lands report reminded Congress of the obdurate realities of the West, it was Powell the systematic geologist who laid out the facts, conjugated their interrelations, and, with relentless objectivity, explained what they meant.

Powell also taught himself what he needed to know to become an anthropologist, and indeed he had to, for the scientific study of humankind was still embryonic when he took it up. Powell played an important role in establishing the discipline of anthropology in the United States, and at the time of his death he was a vice president of the American Anthropological

---

[7] *Exploration of the Colorado River of the West and Its Tributaries,* 153. Opinion is divided today as to whether the Green River is an antecedent stream but not as to the brilliance of this description or of Powell's formulation.

Association and an editor of *American Anthropologist.* He also represented anthropology on the editorial board of the journal *Science,* the respected organ of the American Association for the Advancement of Science, of which he was elected president in 1888—not a bad achievement for a self-taught child of the prairie frontier. Powell came by these distinctions honestly, for in the course of his long career he made particularly enduring contributions to anthropology's subfields of ethnology and linguistics. His interest in these specialties had taken root at least as early as the winter of 1868–69, when he, Emma, and a few others were camped beside the White River in Colorado, west of the continental divide. Through the long snowy weeks of relative inactivity, Powell took great interest in the local Utes who came to visit and trade. As he would do throughout his future relations with Native Americans, he questioned them closely on a wide range of subjects, took note of their habits and stories, and compiled vocabularies of their language.

Most notable of Powell's anthropological achievements was his formulation of the first comprehensive family tree of the languages of North American Indians, the essential structure of which remains accepted today. He created a linguistic taxonomy, patterned on the classificatory methods of the natural sciences, to show the organic relationships between and among languages. Although his efforts in this area generated controversy (Powell denied a competitor access to essentially public information), he ultimately succeeded in devising a durable system and winning the battle to publish his results first.[8]

In addition to his contributions to the physical and social sciences, Powell also became an institution builder. He founded the Bureau of American Ethnology within the Smithsonian Institution, becoming its first director in 1879, a position he retained until his death in 1902. The goal of the bureau was to document and collect all that might reasonably be preserved of the

---

[8] Powell's competitor was Daniel G. Brinton (1837–1899). For a full account of Powell's contribution in this area, see Shaul, "Linguistic Natural History," and Goddard, "The Classification of the Native Languages of North America."

native cultures of the continent: artifacts, mythologies, records of rituals, scraps of languages. The presumption was that many of these cultures would soon perish and that the influences of white America would steadily alter, if not corrupt, those that survived. By commissioning and directing hundreds of field studies through the bureau, Powell and his successors built an ethnological archive that remains today an irreplaceable record of the American past.

Powell, however, is better known as the head of the United States Geological Survey, a post he held from 1881 to 1894. He guided the agency through all but a fraction of its first decade and a half—years that were stormy because of Powell's aggressive leadership and the challenge and creativity of his ideas. The task of advancing those ideas past the realm of the abstract led Powell to yet another role—that of master lobbyist and political infighter. His presentations to Congress became legendary for their persuasiveness, clarity, and command of the facts. Not coincidentally, Powell also distinguished himself as a writer and popularizer. He well knew that the surest political influence sometimes came from going over the heads of Congress and appealing directly to the American people. Nevertheless, in spite of his political savvy and impressive powers of communication, Powell fought nearly all his battles uphill, and in the end victory eluded him.

Extensive geographic study had led Powell to one incontrovertible fact: that the land laws of the United States were fatally ill adapted to the realities of the arid West. Homesteads of 160 acres were too big for family-scale irrigation agriculture and too small for ranches. And neither aggregated homesteads nor any other means of land tenure available under the law suited the needs of communities in the arid lands for collective organization. Powell began to campaign for legislative reform in 1878 with publication of his far-seeing *Report on the Lands of the Arid Region,* and he did not cease to battle for change until he suffered political death at the hands of an angry Congress more than a dozen years later.

His demise resulted from a battle for which he did not choose the time or place. In 1888 Congress charged the U.S. Geological Survey, which Powell headed, with a task he had long urged: The agency was to survey the irrigable lands of the West. This would be a first step toward segregating those lands from the rest of the public domain so that they might be systematically settled and their irrigation potential developed. But Powell's hastily organized Irrigation Survey soon met a storm of controversy. Real estate speculators trailed his survey crews like vultures. They knew that the lands the survey identified as irrigable would quickly increase in value because of the likelihood of their being provided with dams, canals, and other irrigation works, and the speculators filed preemptively on those lands, expecting to sell later to ordinary settlers at a profit. When the Interior Department moved to protect settlers from such exploitation, it took the Draconian action of closing the entire public domain to further settlement until the survey of irrigable lands should be complete. The decision was not Powell's, but its effect fell heaviest on him.

Far more than obstructing speculators, the closure of the public domain inconvenienced almost every individual involved in the ongoing settlement of the West, whether honest or dishonest, rich or poor. It stopped settlement and land trading in their tracks and instantly became a political debacle for the Irrigation Survey, which was widely, if inaccurately, seen as responsible for it. The survey never again sailed in clear weather, and by 1891 Powell's congressional enemies, most of them from the West, had gutted his budget and fatally enfeebled the enterprise that he hoped would lead to the reform he had sought for so long.[9] Ultimately, having lost the culminating battle of his life, Powell withdrew from the Geological Survey and was obliged to turn to other concerns. His life had followed the arc of tragedy, from explorer and hero to national leader and challenger of the status quo, to national repudiation. Having attained a great height, he fell far and hard.

He was a beaten man, but he did not withdraw from engagement in matters of significance. He retained his directorship of the Bureau of Ethnology (renamed the Bureau of American Ethnology in 1894) and continued his

---

[9] See the introduction to Part V, pp. 211–18.

anthropological and sociological investigations. He also indulged a long-standing interest in epistemology and cognition, struggling to define how knowledge arose from the origins of perception. Two years before his death in 1902, he published *Truth and Error,* a long and tangled treatise in which he attempted to integrate all knowledge and all kinds and levels of thought into a unified and coherent whole. At the last, Powell had become a philosopher, albeit with a small audience. Few took him seriously. Even his friends had to admit that much of what he wrote seemed incomprehensible, but, they allowed, he still might have been on to something. Said his friend and colleague Grove Karl Gilbert, "Admitting myself to be of those who fail to understand much of his philosophy, I do not therefore condemn it as worthless, for in other fields of his thought events have proved that he was not visionary but merely in advance of his time."[10]

Powell succumbed to progressive cardiovascular disease at his summer home in Haven, Maine, on September 23, 1902. He was 68 years old. Even in death, Powell wanted to accomplish something, for he left his friends and family with unusual postmortem instructions. He told them that after his death they should arrange to have his brain removed from his skull and examined and weighed.

This grisly procedure was not uncommon among the greater lights of Europe and North America. Based on the erroneous notion that brain weight correlated directly with intelligence, there existed a roster by which great men were ranked according to the heft of their brains. Ivan Turgenev topped the list with a brainy mass of 2,012 grams. The French naturalist George Cuvier came next at 1,830 grams, and from him the list descended through such notables as William Thackeray, E. D. Cope, and Daniel Webster, to the emperor Napoleon III (1,500 grams), whose occupation was succinctly given as "sovereign." Powell evidently wished to be included among this dissected, highbrow elite.

In life he had laid a wager with his long-time friend and ally W. J. McGee

---

[10] Gilbert, "John Wesley Powell" [obituary], 567.

that his brain was larger than McGee's. Presumably the two men had enjoyed ribbing each other on this count. Perhaps they had also taken pleasure in shocking fellow members of the National Geographic Society or the Cosmos Club, both of which Powell helped to found, with their commitment to carry their gruesome rivalry even past the point of knowing its outcome. Powell having died first, it fell to McGee, whom Gifford Pinchot later lauded as "the scientific brains of the Conservation movement," to see that his friend's cadaver was duly dissected and its gray matter examined.[11] In his final illness Powell had likely suffered some wasting of cerebral tissue, which compromised his position on the international roster of heavy thinkers, but with a cerebral mass of 1,488 grams, he still beat McGee, who died in 1912. Today, both of their brains, suitably pickled, remain in jars on a dark shelf in the innards of the Smithsonian Institution in Washington, D.C.

The measurement of Powell's brain stands for more than a weird preoccupation among intellectuals, and it certainly goes beyond Powell's jousting with McGee over the size of his favorite organ. In all of American science and letters, one would be hard pressed to find any figure as dedicated as John Wesley Powell to the synthesis of knowledge across disparate fields. For him there was nothing that was not worth probing, weighing, ranking, and categorizing—not even his own remains. In a kind of intellectual "house that Jack built," Powell attempted to inventory and classify nearly everything he came to know: the watersheds of the American West, the lands within those watersheds and the occupations the land would support, the processes that gave the lands their form, the native tribes that inhabited the lands, the languages spoken by the tribes, the stages of human development embodied both by the tribes and by the Euro-Americans who conquered them, and even the modes and rules of ideation by which the minds of native and conqueror alike operated. No puzzle of existence was

---

[11] Pinchot, *Breaking New Ground*, 359. On Powell's bet with McGee, see Stegner, *Beyond the Hundredth Meridian*, 349. E. A. Spitzka (see note 12) seems also to acknowledge the wager.

too broad to escape his interest; no feature of land or culture was too small to engage his appetite for explanation.

Powell was a difficult man—stubborn, persistent, unyielding—but he was not difficult to befriend. His close associates were many and devoted, and an impressive number of talented professionals worked effectively with him for long periods. Indeed, one of his greatest skills was in nurturing the talents of others. Geologists Grove Karl Gilbert and Clarence Dutton, conservation pioneer W. J. McGee, sociologist Lester Ward, and photographer Jack Hillers are among the most noteworthy of the collaborators whom Powell encouraged and helped to advance to remarkable accomplishments. Difficult he was, but not in the way he treated people. The most difficult thing about Powell was that he was difficult to contain, difficult to pin down, difficult to describe.

A curious sidelight to the examination of his brain lets us glimpse how the people closest to Powell perceived him. Neurologist E. A. Spitzka, who made the examination, sought to make connections between Powell's anatomy and his behavior. Although we may easily dismiss his syllogistic conclusions, we can nevertheless be grateful that he attempted to make them because in his researches, Spitzka collected remembrances about Powell from many of the Major's friends and associates.[12]

One friend mentioned Powell's artistic side: "He wrote much in verse, very little of which has been printed. His tendency was strongly imaginative, notwithstanding that he was addicted to scientific pursuits during so many years."

Another emphasized Powell's powers of empathy: "His great gift of sympathy brought him near the Indian, and he was almost the first to put himself at the Indian point of view, sympathetically."

Several commented on his natural authority and habit of command: "He certainly was a man of suggestion, thus bringing workers to him over whom

---

[12] Spitzka, "A Study of the Brain of the Late Major J. W. Powell," 585–643.

he exerted a marked influence. He was a man qualified to rule and direct."

But in the end, we return where we started, for the most frequently repeated observation had to do with Powell's enormous appetite for empirical information and his compulsion to organize that data within a theoretical framework: "His mind was not satisfied to hold either facts or generalizations without explanation, and his search for explanation extended to the broadest generalizations and most fundamental concepts."

Although the population of the American West has grown exponentially since Powell's time, the region still wrestles with issues of resource allocation and governance in ways that have changed little since Powell's time. We Americans still debate whether western lands should be subject to federal or local control, whether rivers should be dammed or allowed to run free, whether forests should be managed for industrial consumption or for watershed stability. We debate the very character of the lands of the region, and when we have done with that, we debate who by rights should control them and reap their bounty or forgo its reaping. For more than 100 years, the region seems to have faced an agenda of perpetually unfinished business.

A major reason for this incompleteness is that generations of westerners and their eastern counterparts have tried to address the needs of the region by fractions, issue by issue and crisis by crisis. But that was not Powell's way. He saw things whole. He drew his conclusions from the entirety of the puzzle, not piece by piece. And then fought for the whole view, never shying from its implications. The selections that follow are the essence of Powell's legacy. They tell his story, explain his vision, and express his deepest convictions. Few changes have been made to the original texts except to eliminate obvious typographical errors and distracting archaic spellings. Powell's writing, however, constitutes only part of the extraordinary legacy he has left to the people of the American West—and to people everywhere who strive to reconcile the demands of their society with the imperatives of the land. For the rest we must look to his example and to the potential of our land and society to be not two things that argue with each other, but one thing that abides in health.

John Wesley Powell, 1834–1902

1834   Powell is born March 24 in Mount Morris, New York, a village near the Genesee River south–southwest of Rochester.

1838   His family moves to Jackson, Ohio, in the southeast corner of the state.

1846   His family moves to South Grove, in Walworth County, Wisconsin, a few miles north of the Illinois line.

1846–1850 Powell leaves school to work on his family's farm; pursues self-education, especially in natural history.

1851   His family moves to Bonus Prairie, Illinois, less than 20 miles south of South Grove and 15 east of Rockford.

1853–1858 Powell attends Illinois Institute (now Wheaton College), Illinois College (Jacksonville), and Oberlin College, interspersing college terms with spells of country school teaching in Wisconsin and Illinois and extended collecting expeditions along the Mississippi and its tributaries.

1858   Powell begins teaching in Hennipen, Illinois, and becomes secretary of the Illinois State Natural History Society.

1861   Powell enlists as a private in the Twentieth Illinois Volunteer Infantry.

1861   Powell marries his cousin, Emma Dean, November 28.

1862   On April 6 in the Battle of Shiloh a minié ball shatters the bones of Powell's right forearm. Two days later his arm is amputated above the elbow.

1863   Powell returns to active duty in February. Sees action in the Vicksburg campaign.

1864   He commands the artillery of the Seventeenth Corps in the Nashville campaign.

1865   Powell is discharged in January with the rank of major.

1865   He teaches geology and natural history at Illinois Wesleyan University in Bloomington.

1866   Powell resigns from Illinois Wesleyan to take a position at the State Normal University of Illinois, also in Bloomington.

1867   As curator of the Illinois Natural History Society, Powell leads his first expedition, under the auspices of Illinois State Normal University and several other sponsors, to the Rocky Mountains of Colorado.

1868   Powell leads a second expedition to the Rockies, exploring west of the continental divide. Winters at Powell Bottoms, on the White River, in western Colorado.

1869   He leads a party of 10 men, in four boats, down the canyons of the Green and Colorado Rivers, from Green River, Wyoming, to the mouth of the Virgin River in Nevada.

May 24   The expedition departs Green River.

June 8   The *No-Name* is lost at Disaster Falls in Lodore Canyon.

August 13   The expedition breaks camp at the mouth of the Little Colorado River and enters the Grand Canyon.

August 15   The expedition reaches the mouth of Bright Angel Creek (Phantom Ranch).

August 28   The Howland brothers, Oramel and Seneca, and William Dunn depart the expedition at Separation Rapids.

August 30   Powell and the remaining five members of the expedition reach the mouth of the Virgin River in Nevada Territory.

1870   Congress establishes the Geographical and Geological Survey of the Rocky Mountain Region, with Powell in charge.

1870–1879 Powell directs the survey and mapping of the Plateau Province.

1870 He explores portions of the Colorado Plateau with Jacob Hamblin in autumn.

1871–1872 Powell leads a second expedition by boat down the Green and Colorado rivers to the mouth of Kanab Wash.

1877 *Introduction to the Study of Indian Languages* is published.

1878 *Report on the Lands of the Arid Region of the United States, with a More Detailed Account of the Lands of Utah* is published.

1879 Congress establishes the United States Geological Survey, combining the four extant surveys of the West into a single enterprise under the directorship of Clarence King. Congress also establishes the Bureau of Ethnology within the Smithsonian Institution to carry on and extend the ethnographic work of the surveys, and Powell is named its first director.

1879 Powell tours the West for 5 months on behalf of a public lands commission, visiting scores of remote communities.

1881 He succeeds Clarence King as director of the United States Geological Survey while remaining director of the Bureau of Ethnology.

1883–1896 Settlers and speculators file on lands of the public domain at the breakneck rate of 25 million acres per year.

1886–1887 An exceptionally hard winter devastates the range cattle industry of the northern plains. Prolonged drought follows.

1888 Congress instructs the Geological Survey to carry out a broad hydrographic survey of the West, identifying lands necessary for the development of irrigation. The Irrigation Survey is formed.

1889 In August, the Department of Interior orders that the public domain be closed to entry and that the closure be made retroactive to October 2 of the previous year (when funds were appropriated for the aforementioned survey of irrigable lands).

1890 In March, April, and May Powell publishes a series of three articles in *Century Magazine* that set forth his final vision for settling the arid lands and allocating and governing their resources. Meanwhile, Congress rescinds the authority of the Interior Department to withdraw irrigable public land from entry and slashes funding for the Irrigation Survey, effectively killing it.

1891   *Indian Linguistic Families of America North of Mexico* is published.

1894   Powell resigns as director of the Geological Survey.

1895   *Canyons of the Colorado,* a lightly edited version of Powell's report to Congress on the exploration of the Colorado River and its canyons, is published.

1898   *Truth and Error, or the Science of Intellection* is published.

1902   Powell dies at Haven, Maine, September 23.

PART I

Down the Colorado:
Letters from
the Wilderness Post

E ARLY IN THE afternoon of May 24, 1869, a crowd assembled on the banks of the Green River, in Wyoming Territory, just downstream of the bridge where the Union Pacific Railroad crossed the Green. Virtually the entire population of the tiny village of Green River Station had turned out to witness the launch of Major John Wesley Powell's Colorado River Exploring Expedition. There was not a lot to see, but they eyed the stacks of supplies on shore and sometimes chatted with the small group of men stowing the gear into the four boats tied at the river's edge. These boats were to carry the expedition down the Green 350 miles to the little-known Colorado River and thence into either history or oblivion. Opinion was divided as to which, so the atmosphere was more serious than celebratory.

Exactly 2 weeks earlier and many miles to the west, a much larger crowd had gathered beside this same rail line in Promontory, Utah, to see railroaders drive a golden spike joining the tracks of the Central Pacific with those of the Union Pacific. Now that the nation was bound together by the steel ribbons of its first transcontinental railroad, it would not long tolerate blank spots on its map. John Wesley Powell knew this as acutely as anyone, and he aimed to fill the biggest blank still remaining between Atlantic and Pacific shores. Powell was headed into a vast terra incognita that included much of western Colorado, huge portions of southern and eastern Utah, most of northern Arizona, and the northwest corner of New Mexico. Today we call the region the Colorado Plateau. The term arose from the work Powell eventually directs. Over the next two decades he and a succession of capable assistants mapped it, described its properties and native people, photographed and sketched it, and explained it to the American people. Along the way, they recorded some of the greatest adventures in the history of American exploration, especially on this first expedition.

In May 1869, as Powell readied his boats, almost nothing was known of the vast region for which he was bound, save for fragmentary reports from a few mountain men, Indians, and traders. It was known that the Green River, flowing south from Wyoming, joined with the Grand River, which rose in the Colorado Rockies, and that the two rivers, once merged, became a mighty stream called the Colorado that ultimately made its way far to the west and reentered the known world near Mormon settlements at the southern tip of Nevada. But what path the river took to get there, whether it was passable by boat or any other means, the character of the country it passed along the way, the prospects for survival—none of these questions had answers, and that is what attracted John Wesley Powell. The bearded, one-armed major intended to answer them all, and more to boot. In particular, he wanted to divine the geologic history of the land through which the river passed: Of what was it made, how did it form, and what might it offer to the people of the United States?

Powell was likely the smallest of his group. He stood a hair above 5 feet 6 inches and at 35 years of age had scarcely begun to lose the thinness of youth. Exempted by his empty right sleeve from the heavy work of loading provisions, he directed the other men in their tasks. There were 10 in all, including Powell. All but 2 were veterans of the Civil War, and most of them had years of frontier experience, hunting, trapping, or trading on the plains and in the mountains of the West.

Certainly the moodiest of this seasoned and self-reliant group was Walter Henry Powell, the Major's younger brother.[1] Captain Walter Powell had

---

[1] Walter H. Powell should not be confused with Walter Clement "Clem" Powell, the Major's first cousin, who participated in the Major's second Colorado River expedition in 1871.

fought beside his brother at Shiloh, where John Wesley lost his arm. He also served in Sherman's Atlanta campaign until Confederates captured him along with the battery he commanded. As a prisoner, he suffered starvation and fever, then briefly escaped, and when recaptured was found to be quite mad. After confinement in a Confederate hospital, he was returned to the North in a prisoner exchange in 1865 but never entirely regained his mental balance. He remained a brooding presence, quick to anger and undependable except in his loyalty to his brother. Physically powerful, he became a menacing presence as strains within the expedition developed.

John Colton Sumner, 29, had become a mountain man after his service in the war, trapping and outfitting from a base in the Colorado Rockies. He knew the Major well, having helped guide his summer expeditions of 1867 and 1868. Complaining that Powell never paid him properly for his labors (his case for compensation was questionable) and writing long after the expedition had ended, he became one of Powell's severest critics. With some justification, he groused that the Major "gave very scant credit to any of his men. It was all 'Captain Powell and I,' when as a matter of fact Captain Walter Powell was about as worthless a piece of furniture as could be found in a day's journey."[2]

Oramel G. Howland, called O. G., was half a year older than the Major and had come to Colorado from his native Vermont in 1860. Since then he had made his way mainly as a printer and journalist. He was to captain the unhappy *No-Name* in its unsuccessful attempt to pass Disaster Falls, a misfortune that undermined the unity of the expedition. His feeling that the Major wrongly blamed him for the loss of the boat and its precious supplies nourished a sense of injury that later stresses deepened. Without it, Howland and the two who followed him might never have separated from the rest of the group, an act that cost them their lives.

Seneca Howland, 26, was O. G.'s younger brother. Wounded at Gettysburg in 1863 and soon thereafter mustered from the Union Army, he had accepted his brother's invitation to join him in the West. He also followed

---

[2]Cooley, *The Great Unknown*, 194.

his brother when O. G. pulled out from the expedition at the lower end of the Grand Canyon and attempted an ill-fated trek across the Shivwits Plateau to Mormon settlements on the Virgin River.

William H. Dunn joined the Howlands in their overland trek, and his subsequent death on the plateau left his 30-year life shrouded in mystery. Preferring buckskins to cloth, and long, raven hair to any contact with scissors or razor, Dunn self-consciously looked the part of a mountain man. Like Sumner and the Howlands, he had served the Major on his earlier Rocky Mountain expeditions.

Young Billy Hawkins, another former trapper, at times called himself Billy Rhodes or dispensed with surnames altogether and went by "Missouri." He had reason to be flexible with surnames: He was a fugitive from the law. Nevertheless, he proved to be one of the most dependable and relentlessly cheerful members of the expedition.

Like Hawkins, Andy Hall, still a teenager, showed the buoyancy and stamina of youth. Husky and tough, irrepressibly energetic, he had already logged several years as a bullwhacker driving freight wagons on western trails. The Major spotted him on the Green River at the oars of a homemade boat, drew him into a brief conversation, and recruited him on the spot.

George Young Bradley also joined at the last moment. The Major had met him some months earlier at nearby Fort Bridger, where Bradley served as a sergeant major of infantry. Bradley's incipient interest in geology immediately recommended him to Powell, and he desperately wanted to leave the monotony of garrison life. He told Powell he would "gladly explore the river Styx" to do so.[3] Powell sent word to his old friend General U. S. Grant, requesting Bradley's services, and the necessary release arrived just in time to allow the sergeant to join the expedition.

Frank Goodman, an Englishman who claimed to be mad for adventure, had hung around camp a few days, so the men knew him casually. He seemed a likable sort and willing to work, and he begged the Major to take him, even offering to pay to be included. The Major relented, and Good-

---

[3] Darrah, *Powell of the Colorado*, 111.

man became the last of the 10 who shoved their fragile boats into the swift current of the Green.

Four boats made up Powell's little navy. All possessed closed compartments, fore, aft, and at the waist, that were intended to be watertight and to provide buoyancy and safe stowage for provisions and gear. The *Maid of the Canyon, Kitty Clyde's Sister,* and *No-Name* were built of heavy oak and measured 21 feet in length. Carrying them around falls or lining them down rapids, the men would soon curse their bulk and weight. The fourth boat, only 16 feet long and made of pine, was light and maneuverable. This was Powell's pilot boat, which he named the *Emma Dean* in honor of his wife. He himself would ride in it, always going first to scout the way.

The expedition fared well at first, alternately skimming through riffles and rowing stretches of flat water. The men swiftly accustomed themselves to maneuvering the boats and adapted their long-established camp habits to the routines of river travel. And downstream they went. They were out of touch with the rest of the world for 5 full weeks. At the end of June, they arrived at the mouth of the Uinta River, a modest tributary that joins the Colorado from the west. Not far up the creek lay the Uinta Indian Agency, where Powell and others carried letters to be sent back to friends, family, and, not least, to newspapers. They were still hundreds of miles away from encountering the marvels and trials of the Grand Canyon, but they had real news to report: On June 8 one of their boats, the *No-Name,* wrecked spectacularly in crashing whitewater at Disaster Falls, nearly drowning three men and costing the expedition irreplaceable supplies. But no one was lost, and the worst injuries were bruises.

These truths were not half as dramatic as the wholly invented and morbid fiction about the expedition that played in the national press in the early days of July. Thanks to the lurid talk of an opportunistic rogue, the party's correspondence via the "Wilderness Post" found a far more interested audience than Powell might ever have hoped for.

As Powell's letters made slow progress toward their destinations, a character named John A. Risdon materialized from the canyon country and presented himself as the sole survivor of a disaster in which Powell and every other member of the expedition had drowned. Risdon claimed that he had stood on shore and watched as 25 men in a single boat were sucked into a prodigious whirlpool. Through the early days of July, Risdon traded frequent repetition of his tale for meals and passage east to Illinois. But by July 10, Risdon's brief celebrity ran out, and he was jailed for stealing a horse in Springfield. Powell's fame was on the rise, however, as newspapers throughout the country first devoured Risdon's tale of disaster and then joyously rebutted him, printing every word Powell dispatched from the Uinta.

In July 1869 the *New York Times* mentioned Powell in at least 11 separate editions. He was mourned on the 5th, when Risdon's story broke, and again on the 6th when the prevarication was expounded in full detail. On the 7th, Emma Dean Powell brought contradictions to light. On the 9th, a rival and equally untrue account of disaster cropped up. On the 13th the paper ran a story telling "How the Hoax Grew in the Hands of Sensation Mongers." On the 18th the paper reported that letters had been received from the expedition and that all hands were safe. On the 20th it reported that a second expedition under a Captain Samuel Adams was said to be launching from Breckenridge, Colorado, on the Grand River. On the 21st it printed Powell's letter of July 4, written from Flaming Gorge, and on the 23rd it ran his letter to President Edwards of Illinois Normal University, which is reproduced here. On the 25th the *Times* presented O. G. Howland's account of the wreck of the *No-Name*. And on the 26th, an insightful editor of the *Times* observed,

It has been fortunate for the fame of Major Powell, of the Colorado Expedition, that the rascal Risdon propagated the falsehood of the tragical loss of the whole expedition. Up to the circulation of Risdon's hoax, the country had taken hardly any interest in the exploring party, which had started out on its daring and dangerous adventures. But since then, the entire country has taken the deepest interest in the Powell expedition. The newspapers have eagerly printed every item they could get hold of about the party, and every letter or note from any of the members has been universally read. The interest thus aroused will be very encouraging to the explorers, and will be of great help in attracting attention to the region of the Colorado, and securing its thorough exploration. But we don't thank the rascal Risdon for the good that has grown out of his fictions.

The writer's prediction of a "thorough exploration" for the Colorado country would have come true in any event, for the expansion of the nation would demand it. But the fame Powell acquired as a result of the extraordinary perils and achievements of his journey not only ensured the speedy accomplishment of the exploration but also guaranteed that Powell might have the job of carrying it out, if he wanted it. And he would want it badly.

The letters Powell dispatched from the Uinta agency gave the nation its first sense of the drama and daring of his adventure. Two are reproduced here. The first is Powell's brief note to President Edwards of Illinois Normal University confirming the safety of the expedition. Powell was still a member of the faculty at Normal, and the university was helping to sponsor the expedition. The second is a much longer account that he provided to the Chicago *Tribune* of the fateful wreck of the *No-Name*.

—◇—

The Party Has Reached This Point in Safety

———

Camp at the Mouth of Niutah,[1]
June 29, 1869.

Dr. Edwards:

My Dear Sir: The party has reached this point in safety, having run four canyons, of about twenty-five miles in length each, the walls of which were from 2,000 to 2,800 feet hight [sic]. We found falls and dangerous rapids, when we were compelled to make portages of rations, etc., and let the boats down with lines. We wrecked one of our boats and lost about one-third of our supplies and part of the instruments.

The instruments were duplicated, but the loss of rations will compel us to shorten the time for the work. You will perceive an account of our trip more in detail in the Chicago *Tribune,* as I shall send some letters to it for publication. In the wreck I lost my papers, and have to use plant dryers for my letter paper. I have not made a large general collection, but have some fine fossils, a grand geological section, and a good map.

Shall walk to the Niutah Indian Agency, about twenty-five or thirty miles from camp, where I shall mail this letter, and hope to get letters and some news.

———

Powell dispatched this letter from the Uinta Indian Agency to Dr. Richard Edwards, president of the Normal University at Bloomington, Illinois. The *New York Times* reprinted it on July 23, 1869.

[1] Should be *Uintah* or *Uinta.* The misspelling, which is repeated in the body of the letter, may be Powell's or the newspaper editor's. Powell was still new to the geographic vocabulary of the arid lands, which he did more than anyone else to enlarge.

The boats seem to be a success; although filled with water by the waves many times, they never sink. The light cabins attached to the end act well as buoys. The wreck was due to misunderstanding the signal, the Captain of the boat[2] keeping it too far out in the river, and so was not able to land above the falls, but was drifted over.

We shall rest here for eight or ten days, make repairs and dry our rations, which have been wet so many times that they are almost in a spoiling condition, in fact, we have lost nearly half by now by one mishap and another. I have personally enjoyed myself much, the scenery being wild and grand beyond description. All in good health, all in good spirits, and all in high hopes of success. I shall hasten to the Grand and Green,[3] as I am very anxious to make observations on the 7th of August of the eclipse. With earnest wishes for your continued success and prosperity at Normal, I am with great respect yours, cordially,

[Signed] J. W. Powell

---

[2] O. G. Howland.

[3] The junction of the Grand, which is today known as the Colorado, and the Green, which Powell has been descending. The name *Colorado* in Powell's day applied only to the combined river downstream of the junction, which lies southwest of present-day Moab, Utah. Powell had hoped to be in the more open landscape of the junction and out of the confining walls of the river canyon to observe the eclipse, but in fact he made such good time that by August 7 the expedition had passed not only the junction but Cataract Canyon, Glen Canyon, the Crossing of the Fathers, and the future site of Lees Ferry. From deep within Marble Canyon, the Major and his brother Walter climbed for 4 hours to "the summit"—a rim of the canyon—where they awaited the eclipse. Unfortunately,

Clouds come on, and rain falls, and sun and moon are obscured.

Much disappointed, we start on our return to camp, but it is late, and the clouds make the night very dark. Still we feel our way down among the rocks with great care, for two or three hours, though making slow progress indeed. At last we lose our way, and dare proceed no farther. The rain comes down in torrents, and we can find no shelter. We can neither climb up nor go down, and in the darkness dare not move about, but sit and 'weather out' the night. (*Exploration of the Colorado River and Its Canyons*, 237. See note on page 61, this volume.)

Anyone familiar with canyon country will appreciate both Powell's nerve in attempting to descend the wall of Marble Canyon in pitch darkness and his misery in having to weather out the night, trapped on a precarious perch in a driving rain.

The Wreck of the *No-Name*

Colorado River Exploring Expedition,
Echo Park, Mouth of Bear River,[1]
June 18, 1869.

On the 8th our boats entered the Cañon of Lodore—a name suggested by one of the men, and it has been adopted.[2] We soon came to rapids, over which the boats had to be taken with lines. We had a succession of these until noon. I must explain the plan of running these places. The light boat, "Emma Dean," with two good oarsmen and myself explore them, then with a flag I signal the boats to advance, and guide them by signals around dangerous rocks. When we come to rapids filled with boulders, I sometimes find it necessary to walk along the shore for examination. If 'tis thought possible to run, the light boat proceeds. If not, the others are flagged to come on to the head of the dangerous place, and we let down with lines, or make a portage.

As indicated in the previous letter to Edwards, Powell gave a fuller account of the expedition's experiences in letters he sent to the Chicago *Tribune*. This one, published in the *Tribune* of August 20, 1869 (and reprinted in the *Utah Historical Quarterly*, v. 15, 1947), describes the wreck of the *No-Name*, which occurred on June 8 according to most reconstructions of the trip. Powell's published accounts give the date as June 9, but from the mouth of Lodore to the layover at the Yampa, his dates are consistently 1 day behind those of other journal keepers of the expedition. See Cooley, *The Great Unknown*, 124.

Known today as the Yampa River.

According to Darrah (*Powell of the Colorado*, 124) it was Andy Hall who showed unexpected erudition by proposing that the canyon be named for the Cumberland waterfall commemorated in Robert Southey's poem "The Cataract of Lodore" (1820). Southey (1774–1843) was appointed poet laureate of England in 1813.

At the foot of one of these runs, early in the afternoon, I found a place where it would be necessary to make a portage, and, signalling the boats to come down, I walked along the bank to examine the ground for the portage, and left one of the men of my boat to signal the others to land at the right point. I soon saw one of the boats land all right, and felt no more care about them. But five minutes after I heard a shout, and looking around, saw one of the boats coming over the falls. Capt. Howland, of the "No Name," had not seen the signal in time, and the swift current had carried him to the brink. I saw that his going over was inevitable, and turned to save the third boat. In two minutes more I saw that [boat] turn the point and head to shore, and so I went after the boat going over the falls. The first fall was not great, only two or three feet, and we had often run such, but below it continued to tumble down 20 or 30 feet more, in a channel filled with dangerous rocks that broke the waves into whirlpools and beat them into foam. I turned just to see the boat strike a rock and throw the men and cargo out. Still they clung to her sides and clambered in again and saved part of the oars, but she was full of water, and they could not manage her. Still down the river they went, two or three hundred yards to another rocky rapid just as bad, and the boat struck again amidships, and was dashed to pieces. The men were thrown into the river and carried beyond my sight. Very soon I turned the point, and could see a man's head above the waters seemingly washed about by a whirlpool below a rock. This was Frank Goodman clinging to the rock, with a grip on which life depended. As I came opposite I saw Howland trying to go to his aid from the island. He finally got near enough to Frank to reach him by the end of a pole, and letting go of the rock, he grasped it, and was pulled out.[3] Seneca Howland, the captain's brother, was washed farther down the island on to some rocks, and managed to get on shore in safety, excepting

---

[3] Goodman's near drowning at Disaster Falls evidently sated his hunger for adventure. Having also lost all his gear in the wreck of the *No-Name,* 20 days later he bade farewell to the expedition and the river and headed back to civilization by way of the Uinta Indian Agency.

some bad bruises. This seemed a long time, but 'twas quickly done. And now the three men were on the island with a dangerous river on each side, and falls below. The "Emma Dean" was soon got down, and Sumner, one of the men of my boat, started with it for the island. Right skillfully he played his oars, and a few strokes set him at the proper point, and back he brought his cargo of men. We were as glad to shake hands with them as if they had been on a voyage 'round the world and wrecked on a distant coast.[4]

Down the river half a mile we found that the after-cabin of the boat, with part of the bottom ragged and splintered, had floated against a rock, and stranded. There were valuable articles in the cabin, but on examination we concluded that life should not be risked to save them. Of course, the cargo of rations, instruments and clothing was gone. So we went up to the boats and made a camp for the night. No sleep would come to me in those dark hours before the day. Rations, instruments, etc., had been divided among the boats for safety, and we started with duplicates of everything that was a necessity to success; but in the distribution there was one exception, and the barometers were all lost.[5] There was a possi-

---

[4] According to Jack Sumner, Powell greeted Howland with a different kind of fervor:

> This wreck marked the beginning of the many quarrels between Major Powell and O. G. Howland and Bill Dunn. As soon as Howland got out of the boat after the rescue Major Powell angrily demanded of him why he did not land. Howland told him he saw no signals to do anything, and could not see the other boats that had landed until he was drawn into the rapid, when it was too late. I asked Hawkins and Bradley in charge of the other boats if they saw signals to land, and they said no signals were given, but as they saw me turn in they suspected something wrong and followed suit at once. (Cooley, *The Great Unknown*, 56–57).

Sumner, however, wrote his account years after the event and after Powell had achieved the summit of his fame. Powell had already published his narrative of the expedition (1875), which incorporated the text of this letter to the *Tribune* essentially without change, and Sumner's rebuttal of Powell clearly was tinged with jealousy and animosity. Nevertheless, there may be truth in what he says.

[5] Without barometers, Powell could not measure altitude and therefore would not be able to map the region of his explorations accurately.

bility that the barometers were in the cabin lodged against a rock on the island—that was the cabin in which they had been kept. But then how to get to it? And the river was rising—would it be there tomorrow? Could I go out to Salt Lake and get barometers from New York? Well, I thought of many plans before morning, and determined to get them from the island, if they were there.

After breakfast, the men started to make the portage, and I walked down to look at the wreck. There it was still on the island, only carried fifty or sixty feet farther on. A closer examination of the ground showed me it could easily be reached.

That afternoon Sumner and Hall volunteered to take the little boat and go out to the wreck. They started, reached it and out came the barometers. Then the boys set up a shout; I joined them, pleased that they too should be so glad to save the instruments. When the boat landed on our side, I found that the only things saved from the wreck were the three barometers, the package of thermometers and a two gallon keg of whisky.[6] This was what the men were shouting about. They had taken it on board unknown to me, and I am glad they did, for they think it does them good—as they are drenched every day by the melted snow that runs down this river from the summit of the Rocky Mountains and that is a positive good itself.

Three or four days were spent in making this portage, nearly a mile long, and getting down the rapids that followed in quick succession. On the night of the 12th, we camped in a beautiful grove of box elders on the left bank, and here we remained two days to dry our rations, which were in a spoiling condition. A rest, too, was needed.

I must not forget to mention that we found the wreck of a boat near our own, that had been carried above high-water mark, and with it the lid of a bake-oven, an old tin plate and other things, showing that some one else had been wrecked there and camped in the cañon after the disaster. This,

---

[6] In later accounts Powell enlarged the keg to three gallons, and Sumner said it held ten (Cooley, *The Great Unknown*, 57). It must have lasted longer than anyone expected.

think, confirms the story of an attempt to run the cañon, some years ago, that has been mentioned before.[7]

On the 14th Howland and I climbed the walls of the cañon, on the west side, to an altitude of two thousand feet. On looking over to the west we saw a park[8] five or six miles wide and twenty-five or thirty long. The cliff formed a wall between the cañon and the park, for it was eight hundred feet down the west side to the valley. A creek came winding down the park twelve hundred feet above the river and cutting the wall by a cañon; it at last plunged a thousand feet by a broken cascade into the river below. The day after, while we made another portage, a peak on the east side was climbed by two of the men and found to be twenty-seven hundred feet high. On each side of the river, at this point, a vast amphitheatre has been cut out, with deep, dark alcoves and massive buttresses, and in these alcoves grow beautiful mosses and ferns.

While the men were letting the boats down the rapids, the "Maid of the Cañon" got her bow out into the current too far and tore away from them, and the second boat was gone. So it seemed; but she stopped a couple of miles below in an eddy, and we followed close after. She was caught—damaged slightly by a thump or two on the rocks.

Another day was spent on the waves, among the rocks, and we came down to Alcove Creek, and made an early halt for the night. With Howland,

---

[7] In a letter dated June 2, 1869 that appeared in the Chicago *Tribune* on July 19 Powell had written,

> On a rock, by which our trail ran, was written "Ashley," with a date, one figure of which was obscure—some thinking it was 1825, others 1855. I had been told by old mountaineers of a party of men starting down the river, and Ashley was mentioned as one; and the story runs that the boat was swamped, and some of the party drowned in the cañon below. This word "Ashley" is a warning to us and we resolve to use great caution.

William Ashley (1778–1838) was a leading figure in the American fur trade. In April and May 1825 he and a party of five men descended the Green River in bullboats (shallow-draft tubs sheathed with the hides of bull bison). In their 31 days on the river they passed all the country Powell had thus far seen, including the canyon of Lodore. They left the river just above the mouth of Desolation Canyon with no loss of life. (See Goetzmann, *Exploration and Empire*, chapter 4.)

[8] An extensive grassland.

I went to explore the stream, a little mountain brook, coming down from the heights into an alcove filled with luxuriant vegetation.

The camp was made by a group of cedars on one side and a mass of dead willows on the other.

While I was away, a whirlwind came and scattered the fire[9] among the dead willows and cedar spray, and soon there was a conflagration. The men rushed for the boats, leaving all behind that they could not carry at first. Even then, they got their clothes burned and hair singed, and Bradley got his ears scorched.

The cook filled his arms with the mess kit, and jumping on to the boat, stumbled and threw it overboard, and his load was lost. Our plates are gone, our spoons are gone, our knives and forks are gone; "Water ketch 'em," "H-e-a-p ketch 'em."[10] There are yet some tin cups, basins and camp kettles, and we do just as well as ever.

When on the boats the men had to cut loose, or the overhanging willows would have set the fleet on fire, and loose on the stream they had to go down, for they were just at the head of rapids that carried them nearly a mile where I found them. This morning we came down to this point. This had been a chapter of disasters and toils, but the Cañon of Lodore was not devoid of scenic interest. 'Twas grand beyond the power of pen to tell. Its waters poured unceasingly from the hour we entered it until we landed here. No quiet in all that time; but its walls and cliffs, its peaks and crags, its amphitheaters and alcoves told a story that I hear yet, and shall hear, and shall hear, of beauty and grandeur.

---

[9] The cooking fire.

[10] Powell's previously mentioned letter of June 2 includes the following passage. The Indian mentioned probably was Ute.

Last spring I had a conversation with an old Indian who told me of one of his tribe making the attempt to run this cañon in a canoe with his wife and little boy. "The rocks," he said, holding his hands above his head, his arms vertical, and looking between them to the heavens, "the rocks h-e-a-p-h-e-a-p high. The water goes boo-wough, boo-wough! Water pony (the boat) h-e-a-p buck! Water ketch 'em! No see 'em Ingin any more! No see 'em papoose any more!"

Sunday, June 20, 1869.

At the point where the Bear, or with greater correctness the Yampa River enters the Green, the river runs along a rock about 700 feet high and a mile long, then turns sharply around to the right and runs back parallel to its former course for another mile, with the opposite sides of this long narrow rock for its bank. On the east side of the river, opposite the rock and below the Yampa, is a little park just large enough for a farm.[11]

The river has worn out hollow domes in this sandstone rock, and stand-

---

[11] This is Echo Park, which lies a short distance upstream of the Colorado's entry into Utah from Colorado. According to Powell, an event occurred here that illustrates how little he allowed the loss of his arm to curb his physical activity—and how much he should have. The event, which Bradley, who rescued Powell, recorded as taking place on July 8 at a location well down Desolation Canyon (Cooley, *The Great Unknown*, p. 96), also includes the most noteworthy use of long underwear in the annals of western American exploration. As Major Powell tells the tale,

We have named the long peninsular rock on the other side Echo Rock. Desiring to climb it, Bradley and I take the little boat and pull up stream as far as possible, for it cannot be climbed directly opposite. . . .

We start up a gulch; then pass to the left on a bench along the wall; then up again over broken rocks; then we reach more benches, along which we walk, until we find more broken rocks and crevices, by which we climb; still up, until we have ascended 600 or 800 feet, when we are met by a sheer precipice. Looking about, we find a place where it seems possible to climb. I go ahead; Bradley hands the barometer to me, and follows. So we proceed, stage by stage, until we are nearly to the summit. Here, by making a spring, I gain a foothold in a little crevice, and grasp an angle of the rock overhead. I find I can get up no farther and cannot step back, for I dare not let go with my hand and cannot reach foothold below without. I call to Bradley for help. He finds a way by which he can get to the top of the rock over my head, but cannot reach me. Then he looks around for some stick or limb of a tree, but finds none. Then he suggests that he would better help me with the barometer case, but I fear I cannot hold on to it. The moment is critical. Standing on my toes, my muscles begin to tremble. It is sixty or eighty feet to the foot of the precipice. If I lose my hold I shall fall to the bottom and then perhaps roll over the bench and tumble still farther down the cliff. At this instant it occurs to Bradley to take off his drawers, which he does, and swings them down to me. I hug close to the rock, let go with my hand, seize the dangling legs, and with his assistance am enabled to gain the top.

Then we walk out on the peninsular rock, make the necessary observations for determining its altitude above camp, and return, finding an easy way down. (*Exploration of the Colorado River and Its Canyons*, 168–69. See note page 61, this volume.)

ing opposite, your words are repeated with a strange clearness but soft-
ened, mellow tone. Conversation in a loud key is transformed into magical
music. You can hardly believe that 'tis the echo of your own voice. In some
places two or three echoes come back, in others the echoes themselves are
repeated, passing forth and back across the river, for there is another rock
making the eastern wall of the little park. To hear these echoes well, you
must shout. Some thought they could count 10 or 12 echoes. To me they
seemed to rapidly vanish in multiplicity, auditory perspective, or peraudi-
tory, like the telegraph poles on an outstretched prairie. I observed this
same phenomenon once before among the cliffs near Long's Peak,[12] and
was delighted to meet with it again.

J. W. Powell

---

[12] Long's Peak rises more than 14,000 feet in the heart of what is today Rocky Mountain
National Park, northwest of Denver. Powell led a group of seven men, including Walter Powell and Jack Sumner, in the first recorded ascent of the mountain, August 20–23, 1868.

PART II

Voyage into the
Great Unknown

N OTHING REVEALS THE character of a man or woman so much as the individual's conduct under conditions of privation and duress. This is partly what makes the literature of exploration so fascinating: Not only is the reader treated to the vicarious experience of new lands and cultures, but most narratives of exploration also invite the reader on intense inner journeys.

Certainly John Wesley Powell and his men were tested in their descent of the Colorado River. They started as an expedition of 10 men in four boats carrying what they thought to be provisions for 10 months. Ninety-eight days and nearly 900 twisting river miles later, only 6 of them remained. One of their comrades, Frank Goodman, walked out to safety at the Uinta agency after only 5 weeks on the river. His near drowning at Disaster Falls and the loss of all his kit had sapped his desire for further adventure. Three others, the brothers Oramel and Seneca Howland and William Dunn, split off from the rest of the expedition on the 96th day. They climbed from the depths of the Grand Canyon and struck out across the wilderness of the Shivwits Plateau, hoping to reach the Mormon settlements of the lower Virgin River. They were never heard from again, and the manner of their death, discussed in Part III of this book, remains an enduring mystery of the Southwest.

The expedition's tiny flotilla of boats had likewise dwindled. The

No-Name broke to pieces at Disaster Falls; subsequently, Powell's own *Emma Dean* was abandoned when the departure of the Howlands and Dunn left the expedition undermanned; only the *Maid* and the *Sister* made it down the last stretch of the Grand Canyon and reached safety at the mouth of the Virgin River.

The deliverance of what was left of the expedition came just in time, for the once-plentiful provisions were nearly exhausted. At the end, the men possessed but 10 pounds of marginally edible flour and 15 pounds of dried apples. They still had, however, 70 or 80 pounds of coffee, and if caffeine could have substituted for calories, they might also have had a fuller supply of something even more precious: peace of mind. But in the last weeks of their exploration, nothing was in shorter supply.

The prospect of starvation competed with other worries. What few clothes they had left were in tatters, their rotted canvas tents afforded no protection from the elements, and their boats had been battered nearly to splinters and recaulked and recarpentered again and again. What tortured them most, however, especially through their last weeks in the "gloomy prison" of the Grand Canyon, was uncertainty of what lay ahead: "Ever before us has been an unknown danger, heavier than immediate peril," wrote Powell. Every bend of the ever-bending river promised a new roar of cataracts and rapids, another set of trials, another dance with death. Neither Powell nor any of his men was given much to introspection. They were all brave men, and all had been seasoned by experience, most of them in the Civil War. But the Grand Canyon tested them as nothing else could have done. Jack Sumner may have spoken for them all when, reflecting on a day in Grand Canyon whitewater, he commented, "I have been in a cavalry charge, charged the batteries, and stood by the guns to repel a charge. But never before did my sand run so low. In fact, it all ran out, but as I had to have some more grit, I borrowed it from the other boys and got along all right after that."[1]

Powell's account of the expedition, first printed as a government report

---

[1]Cooley, *The Great Unknown*, 159. See also Selection 3, note 4.

under the title *The Exploration of the Colorado River of the West and Its Tributaries,* may be read on a number of levels. As an adventure story, it is one of the best the American West has produced, and the West has produced many great ones. It would be hard to find in any literature a passage more stirring than that found in Powell's account of August 13, 1869 (at which point Selection 3 commences):

> We have an unknown distance yet to run, an unknown river to explore. What falls there are, we know not; what rocks beset the channel, we know not; what walls rise over the river, we know not. Ah, well! we may conjecture many things. The men talk as cheerfully as ever; jests are bandied about freely this morning; but to me the cheer is somber and the jests are ghastly.

Powell's narrative may also be understood as a geological investigation. It has the merit not only of posing superb questions about how the spectacular landscapes of the Colorado Plateau took shape but of providing enduring answers in vivid prose. No one before Powell had explained in such detail the work of erosion in carving the face of a particular region of the earth. Through the course of the *Exploration,* one sees Powell continually at work, identifying the parent materials of the canyons and mesas, analyzing their relationships, and deducing the likely interactions of water and stone through time. Because his enthusiasm for these queries is never absent, what he gives us, in the end, is a kind of rhapsody on rock.

But his interests are by no means limited to matters geological. They flow and spread into all aspects of the canyon world, with the result that the *Exploration* may be read as a great achievement in nature writing. It more than evokes the river and its canyons faithfully; it inspirits them and brings them alive as vigorously as words are capable of doing. Our sense of the grandeur and power of the Colorado River is born with Powell, who first described it with both passion and detail. The same is true of his treatment of the land through which the river passes. Not only did Powell find the canyon to be grand, he *made it grand* in the minds and imaginations of Americans from his time to our own. Even today, probably few groups float the Colorado through the Grand Canyon without a copy of Powell's *Exploration* stowed in a river bag.

And the *Exploration* may also be read as a study of group character and behavior. Each of the men—Sumner, the Howlands, Dunn, Bradley, Hawkins, Hall, Walter Powell, and the Major—has his own trajectory through journey, and their interactions are wonderfully complex. The dynamics of their friendships, grudges, and, above all, their cooperation under conditions of greatest adversity are perhaps best appreciated by reading Powell's narrative in conjunction with the journals of other expedition members,[2] but Powell's *Exploration,* taken alone, easily remains the most powerful and vivid of all the accounts. It evidences depth, intrigue, and a dramatic structure that climaxes with the fatal breakup of the group.

Ultimately Powell's narrative may be read as autobiography. In it one finds revealed every feature of the Major's character—his strengths, weaknesses, eccentricities, and contradictions.

The strengths are clear. A list of them must begin with Powell's rock-hard sense of purpose, his unwavering focus and self-discipline, and his seemingly inexhaustible well of courage. Most people, having lost an arm and hazarding no criticism from the rest of the world, would lead the balance of their lives with somewhat diminished daring. Powell, by contrast, plunged into the wildest blank spot on the map of North America outside the far north and rode a whitewater roller coaster from one side of the blank to the other, never deviating from his original intent. He braved stupendous rapids again and again, more than once being thrown into the drink and swimming for his life with only one arm. Nor did he allow his disability to curb his appetite for rock climbing, although at times, as mentioned earlier, it placed him in extraordinary danger.

Powell's weaknesses also emerge from his account, mostly in what he does not say. He seems to have been largely insensitive to the morale of his men. He was in charge, they were to work for him, and he left things at that. One may conjecture that had he done otherwise—had he worked harder at

---

[2]See Cooley's fascinating concordance of multiple accounts in *The Great Unknown.*

maintaining consensus within his small group, for instance—the departure and consequent deaths of the two Howlands and William Dunn might have been avoided. But by the same token, a more democratically organized expedition might have abandoned the canyon and failed to achieve as much.

Powell's men had a vivid sense of what they believed to be the Major's other weaknesses. One was his conservatism, which contrasts interestingly with his personal recklessness. Time and again, he declined to run the rapids of the Colorado and ordered his men to portage around them or to "line" the boats down, which meant to lower them by ropes carefully and slowly through the rapids. In either case, the work was difficult and laborious, and it was a task from which the Major generally was exempted because of his disability. Off he would go to geologize while the men sweated and strained with the boats.

In the years after the expedition another weakness became evident: The Major made no effort to share credit for the expedition's success. Jack Sumner, whom fate stung more than once with reversals of fortune, became especially bitter about Powell's reluctance to credit his contributions and those of others. Powell's obtuseness in this matter becomes all the more difficult to fathom as one considers the manner in which his record of the expedition was published and the importance to that document of a second but entirely unmentioned Colorado River expedition.

Powell had received limited government assistance in organizing and provisioning his initial expedition. (Once again, his friendship with General Grant proved helpful.) In 1871 Congress awarded him $10,000 for further investigations, including a second descent of the river, but once that expedition was complete, Powell still lacked what he most needed: secure funding for long-term study of the Colorado Plateau and its people. He had chosen his path. His former teaching career no longer held interest. He had determined that his life's work lay in the West—and in Washington, the seat of power and policy.

Powell well knew how to build support for the work he wanted to continue. In June 1874 he submitted a report on his explorations to Joseph Henry, secretary of the Smithsonian Institution. The report, which included Powell's reworked journal of the 1869 river exploration, an account of further journeys by land in 1870, and a discussion of his geological findings, was in turn duly submitted to Congress in 1875. Titled *Exploration of the Colorado River of the West and Its Tributaries,* it entered in the annals of government as House Miscellaneous Document 300 of the 43rd Congress, 1st Session. Obviously, if the expedition occurred in 1869 and the full account was not forthcoming until 1875, Powell had plenty of time to reflect on the meaning of his journey and polish his account of it. He had enough time to descend the river a second time, as we know, and to explore the region by horseback on a number of extended trips. These additional reconnaissances inform Powell's account of his narrative of 1869, yet he does not acknowledge their contribution. Most astonishing, he fails to mention the names of the men who risked life and limb to help him descend the Colorado in his second expedition of 1871–72.

A world of significance inhabits this decision. From the start Powell had addressed his report to a popular as well as government audience. Portions of it appeared in serial form in *Scribner's Monthly,* also in 1875, when the ink of the government submission was hardly dry. The report eventually was published as a popular book in 1895 and has only intermittently been out of print since then. From a literary point of view, his omission of the second river expedition made sense, for the second descent of the river lacked the dramatic structure of the first. Unlike the earlier foray, it was not a penetration of the Great Unknown. Powell knew where he was going, and he wisely arranged for periodic resupply along the way. (Even so, the second expedition suffered at times from want of food.) The second trip was also discontinuous. The Major departed the expedition at the mouth of the Yampa, leaving the others to descend Desolation Canyon on their own. After rejoining the boats for the descent of Cataract Canyon, Powell left again for another land excursion at the Crossing of the Fathers, a site rich in history that now lies beneath the waters of Lake Powell. The expedition continued without him to Lees Ferry, where river operations were suspended for the winter.

Still less acceptably dramatic was the fact that Powell terminated the second river expedition nearly 180 river miles short of their destination. After the winter layover of 1871–72, Powell and his men resumed their voyage, but the river was higher and wilder than ever. They weathered their descent of Marble Canyon, but in the upper reaches of Grand Canyon, the river ran so fast and powerful that boats repeatedly flipped and several men, including Powell, nearly drowned. Their experience was sufficiently daunting that Powell elected to abandon the expedition at the mouth of Kanab Creek, only about 45 miles downstream of Bright Angel Creek and the heart of the inner gorge.

The decision was remarkable, especially considering that it was the lower part of the canyon that Powell most needed to map, given the neglect of technical matters during the last grim days of the 1869 expedition. Rumors had reached the explorers that the Shivwits were angry and would attack as they passed the lower canyons, but it is hard to believe that such news would have dissuaded Powell. He and his men were invulnerable as long as they were on the water. They were well armed, and they needed only to choose secure campsites to pass safely through Shivwits territory. The problem was the river: Ahead lay Lava Falls, Separation Rapid, and Lava Cliff Rapid, where, as Selection 3 relates, the first expedition had nearly met disaster. With the river high and fast and the boats unmanageable, Powell backed down from a challenge for what may have been the first and only time in his life. It was probably a wise choice. Had he done differently, the remainder of his life might well have been counted in days.

All these considerations—the aborted trip, the convoluted itinerary, the lack of absolute newness—justify in literary terms Powell's decision not to complicate the narrative of the first expedition with a detailed account of the second. Nevertheless, it is hard to fathom why he omitted all mention of the men who served and suffered with him in that effort. Some would serve him loyally for years. Almon Harris Thompson, the Major's brother-in-law, was the expedition's chief topographer and acquitted himself far better than many other brothers-in-law in that nepotistic age. Three men, all with connections to schools back in Indiana, assisted Thompson: Walter Graves, F. M. Bishop, and S. V. Jones. Powell brought along a photographer, E. O. Beaman, who performed poorly and was briefly replaced by James Fennemore, who in turn was replaced by Jack Hillers, a teamster by background who received his on-the-job training in the canyons of the Colorado and emerged as one of the early West's premier photographers. Frederick Dellenbaugh, who like Hillers also distinguished himself in later years, was the expedition's official artist and also helped Thompson with topography. So did Derby Johnson, who joined in midvoyage. The rest of the roster included Andrew Hattan, the cook, the Major's young cousin, Clem Powell, and Frank Richardson, who deserted the expedition while it was still in its early stages.

To defuse criticism for having given short shrift to the events and men of the second expedition, Powell later protested that he had been interested mainly in the scientific results of his explorations and that he produced the *Exploration* only at the insistence of others who told him he was obliged to satisfy the public's curiosity.[3] Such a stance is at best disingenuous. Even before he dispatched his first letters to the Chicago *Tribune* via wilderness post, Powell knew well that popular adulation would help him advance his personal fortunes and professional interests. Years later, as he fought battle after battle in the halls of government, he made good use of the leverage that his status as public hero gave him. It is hard to avoid the conclusion that in his report to Congress Powell hogged the spotlight and at times fudged the truth for the sake of ambition.[4] Considering how far ambition carried him, perhaps we should not be surprised. Many fiercely driven men and women have been guilty of worse. The popular account of the *Exploration,* however, did not appear until 1895, when it was eagerly received. By then, almost all his public battles had been fought. One wonders whether it might then have been easier for him to give credit where it was due, but he did not. Perhaps he did not want to appear inconsistent; perhaps he feared that mention of the others would intrude on the way history remembered

---

[3]See especially his preface to the 1895 Flood and Vincent edition of *Canyons of the Colorado,* as the *Exploration* was then titled.

[4]As to falsification, see Goetzmann, *Exploration and Empire,* 559.

him; perhaps, with selfish disregard for the facts, he did not view their contributions as significant. We cannot know; we can only speculate.

What remains still more significant about Powell and more meaningful in terms of his historical legacy is that he never fudged or lied about the lessons he drew from his work. He did not equivocate about what he saw to be the true character of western lands—not for the ease and benefit of his patrons, not to avoid the blandishments of his enemies, not even to endear himself to the public. He never sweetened the pill. He may have too much coveted the glory of his expeditions, but he did not flinch from pursuing the implications of what he learned or from accepting the consequences of what he did and said. He was as obstinate and as hard to move as a fireplug.

Where Powell was concerned, stubbornness was a survival trait. It helped carry him through Shiloh and all its grievous aftermath, and it accounted for much of his success and the success of his ideas years later in the halls of government in Washington. It also stood him in good stead in the defining experience of his life, his first descent of the Colorado through Grand Canyon in 1869. Put politely, stubbornness becomes perseverance, and Powell's *Exploration,* while being many things, is above all a study in perseverance.

———◇———

Through the Grand Canyon
from the Little Colorado to the Virgin River

———

We join Powell and his beleaguered party in 1869 at the confluence of the Little Colorado River and the main stem of the big Colorado. The expedition has survived Marble Canyon, which Powell so named because of its immense, red-stained limestone walls. The expedition is about to enter yet another, even deeper canyon, which Powell also named. He called it Grand.

———

August 13. We are now ready to start on our way down the Great Unknown. Our boats, tied to a common stake, chafe each other as they are tossed by the fretful river. They ride high and buoyant, for their loads are lighter than we could desire. We have but a month's rations remaining. The flour has been resifted through the mosquito-net sieve; the spoiled bacon has been dried and the worst of it boiled; the few pounds of dried apples have been spread in the sun and reshrunken to their normal bulk. The sugar has all melted and gone on its way down the river. But we have a large sack of coffee. The lightening of the boats has this advantage: they will ride the waves better and we shall have but little to carry when we make a portage.

We are three quarters of a mile in the depths of the earth, and the great river shrinks into insignificance as it dashes its angry waves against the walls and cliffs that rise to the world above; the waves are but puny ripples,

<hr>

This excerpt is taken from Powell, *The Exploration of the Colorado River and Its Canyons* (New York: Dover, 1961), Chapter XI. The Dover edition is a facsimile of the 1895 Flood and Vincent volume published as *Canyons of the Colorado*, which in turn is an expanded version of the 1875 report to Congress.

<hr>

and we but pigmies, running up and down the sands or lost among the boulders.

We have an unknown distance yet to run, an unknown river to explore. What falls there are, we know not; what rocks beset the channel, we know not; what walls rise over the river, we know not. Ah, well! we may conjecture many things. The men talk as cheerfully as ever; jests are bandied about freely this morning; but to me the cheer is somber and the jests are ghastly.

With some eagerness and some anxiety and some misgiving we enter the canyon below and are carried along by the swift water through walls which rise from its very edge. They have the same structure that we noticed yesterday—tiers of irregular shelves below, and, above these, steep slopes to the foot of marble cliffs. We run six miles in a little more than half an hour and emerge into a more open portion of the canyon, where high hills and ledges of rock intervene between the river and the distant walls. Just at the head of this open place the river runs across a dike; that is, a fissure in the rocks, open to depths below, was filled with eruptive matter, and this on cooling was harder than the rocks through which the crevice was made, and when these were washed away the harder volcanic matter remained as a wall, and the river has cut a gateway through it several hundred feet high and as many wide. As it crosses the wall, there is a fall below and a bad rapid, filled with boulders of trap;[1] so we stop to make a portage. Then on we go, gliding by hills and ledges, with distant walls in view; sweeping past sharp angles of rock; stopping at a few points to examine rapids, which we find can be run, until we have made another five miles, when we land for dinner.

Then we let down with lines over a long rapid and start again. Once more the walls close in, and we find ourselves in a narrow gorge, the water again filling the channel and being very swift. With great care and constant watchfulness we proceed, making about four miles this afternoon, and camp in a cave.

August 14. At daybreak we walk down the bank of the river, on a little

---

[1] The trap to which Powell refers is basalt. The rapid is known today as Lava Canyon Rapid or Chuar Rapid.

sandy beach, to take a view of a new feature in the canyon. Heretofore hard rocks have given us bad river; soft rocks, smooth water; and a series of rocks harder than any we have experienced sets in. The river enters the gneiss! We can see but a little way into the granite gorge, but it looks threatening.[2]

After breakfast we enter on the waves. At the very introduction it inspires awe. The canyon is narrower than we have ever before seen it; the water is swifter; there are but few broken rocks in the channel; but the walls are set, on either side, with pinnacles and crags; and sharp, angular buttresses, bristling with wind- and wave-polished spires, extend far out into the river.

Ledges of rock jut into the stream, their tops sometimes just below the surface, sometimes rising a few or many feet above; and island ledges and island pinnacles and island towers break the swift course of the stream into chutes and eddies and whirlpools. We soon reach a place where a creek comes in from the left, and, just below, the channel is choked with boulders, which have washed down this lateral canyon and formed a dam, over which there is a fall of 30 or 40 feet; but on the boulders foothold can be had, and we make a portage.[3] Three more such dams are found. Over one we make a portage; at the other two are chutes through which we can run.

As we proceed the granite rises higher, until nearly a thousand feet of the lower part of the walls are composed of this rock.

About eleven o'clock we hear a great roar ahead, and approach it very cautiously. The sound grows louder and louder as we run, and at last we find ourselves above a long, broken fall, with ledges and pinnacles of rock obstructing

---

[2] The expedition is about to enter the inner gorge of the Grand Canyon, crossing the geological boundary between Cambrian and Precambrian formations known as the Great Unconformity. An unconformity is a gap in the geological record—in this case, a period from about 600 million years to more than 800 million years before the present, during which no rocks were laid down. The ancient Precambrian granites and schists of the inner gorge date back a billion years and more.

[3] 75 Mile Creek comes in from river left, just above Nevills Rapid. Hance Rapid, its boulder fields fed by creeks from either side, is a mile below it. Hance is one of the most difficult rapids in the canyon and is sufficiently long that Powell might have thought of it not as one rapid but several.

the river.[4] There is a descent of perhaps 75 or 80 feet in a third of a mile, and the rushing waters break into great waves on the rocks, and lash themselves into a mad, white foam. We can land just above, but there is no foothold on either side by which we can make a portage. It is nearly a thousand feet to the top of the granite; so it will be impossible to carry our boats around, though we can climb to the summit up a side gulch and, passing along a mile or two, descend to the river. This we find on examination; but such a portage would be impracticable for us, and we must run the rapid or abandon the river. There is no hesitation. We step into our boats, push off, and away we go, first on smooth but swift water, then we strike a glassy wave and ride to its top, down again into the trough, up again on a higher wave, and down and up on waves higher and still higher until we strike one just as it curls back, and a breaker rolls over our little boat. Still on we speed, shooting past projecting rocks, till the little boat is caught in a whirlpool and spun round several times. At last we pull out again into the stream. And now the other boats have passed us. The open compartment of the "Emma Dean" is filled with water and every breaker rolls over us. Hurled back from a rock, now on this side, now on that, we are carried into an eddy, in which we struggle for a few minutes, and are then out again, the breakers still rolling over us. Our boat is unmanageable, but she cannot sink, and we drift down another hundred yards through breakers—how, we scarcely know. We find the other boats have turned into an eddy at the foot of the fall and are waiting to catch us as we come, for the men have seen that our boat is swamped. They push out as we come near and pull us in against the wall. Our boat bailed, on we go again.

The walls now are more than a mile in height—a vertical distance difficult to appreciate. Stand on the south steps of the Treasury building in Washington and look down Pennsylvania Avenue to the Capitol; measure

---

[4] Sockdolager Rapid. The Powell expedition is said to have bestowed this name. A sockdolager, in the vernacular of the day, was a roundhouse, knockout punch. The events of this trying day prompted Sumner to compose the comments quoted in the Part II introduction: "Never before did my sand run so low. In fact, it all ran out, but as I had to have some more grit, I borrowed it from the other boys and got along all right after that" (Cooley, *The Great Unknown*, 159).

this distance overhead, and imagine cliffs to extend to that altitude, and you will understand what is meant; or stand at Canal Street in New York and look up Broadway to Grace Church, and you have about the distance; or stand at Lake Street bridge in Chicago and look down to the Central Depot, and you have it again.

A thousand feet of this is up through granite crags; then steep slopes and perpendicular cliffs rise one above another to the summit. The gorge is black and narrow below, red and gray and flaring above, with crags and angular projections on the walls, which, cut in many places by side canyons, seem to be a vast wilderness of rocks. Down in these grand, gloomy depths we glide, ever listening, for the mad waters keep up their roar; ever watching, ever peering ahead, for the narrow canyon is winding and the river is closed in so that we can see but a few hundred yards, and what there may be below we know not; so we listen for falls and watch for rocks, stopping now and then in the bay of a recess to admire the gigantic scenery; and ever as we go there is some new pinnacle or tower, some crag or peak, some distant view of the upper plateau, some strangely shaped rock, or some deep, narrow side canyon.

Then we come to another broken fall, which appears more difficult than the one we ran this morning.[5] A small creek comes in on the right, and the first fall of the water is over boulders, which have been carried down by this lateral stream. We land at its mouth and stop for an hour or two to examine the fall. It seems possible to let down with lines, at least a part of the way, from point to point, along the right-hand wall.

So we make a portage over the first rocks and find footing on some boulders below. Then we let down one of the boats to the end of her line, when she reaches a corner of the projecting rock, to which one of the men clings and steadies her while I examine an eddy below. I think we can pass the other boats down by us and catch them in the eddy. This is soon done, and the men in the boats in the eddy pull us to their side. On the shore of this little eddy there is about two feet of gravel beach above the water. Standing on

---

[5] Grapevine Rapid.

this beach, some of the men take the line of the little boat and let it drift down against another projecting angle. Here is a little shelf, on which a man from my boat climbs, and a shorter line is passed to him, and he fastens the boat to the side of the cliff; then the second one is let down, bringing the line of the third. When the second boat is tied up, the two men standing on the beach above spring into the last boat, which is pulled up alongside of ours; then we let down the boats for 25 or 30 yards by walking along the shelf, landing them again in the mouth of a side canyon. Just below this there is another pile of boulders, over which we make another portage. From the foot of these rocks we can climb to another shelf, 40 or 50 feet above the water.

On this bench we camp for the night. It is raining hard, and we have no shelter, but find a few sticks which have lodged in the rocks, and kindle a fire and have supper. We sit on the rocks all night, wrapped in our ponchos, getting what sleep we can.

August 15. This morning we find we can let down for 300 or 400 yards, and it is managed in this way: we pass along the wall by climbing from projecting point to point, sometimes near the water's edge, at other places 50 or 60 feet above, and hold the boat with a line while two men remain aboard and prevent her from being dashed against the rocks and keep the line from getting caught on the wall. In two hours we have brought them all down, as far as it is possible, in this way. A few yards below, the river strikes with great violence against a projecting rock and our boats are pulled up in a little bay above. We must now manage to pull out of this and clear the point below. The little boat is held by the bow obliquely up the stream. We jump in and pull out only a few strokes, and sweep clear of the dangerous rock. The other boats follow in the same manner and the rapid is passed.

It is not easy to describe the labor of such navigation. We must prevent the waves from dashing the boats against the cliffs. Sometimes, where the river is swift, we must put a bight of rope about a rock, to prevent the boat from being snatched from us by a wave; but where the plunge is too great or the chute too swift, we must let her leap and catch her below or the undertow will drag her under the falling water and sink her. Where we wish to run her out a little way from shore through a channel between rocks, we first throw in little sticks of driftwood and watch their course, to see where we must steer so that she will pass the channel in safety. And so we hold, and let go, and pull, and lift, and ward[6]—among rocks, around rocks, and over rocks.

And now we go on through this solemn, mysterious way. The river is very deep, the canyon very narrow, and still obstructed, so that there is no steady flow of the stream; but the waters reel and roll and boil, and we are scarcely able to determine where we can go. Now the boat is carried to the right, perhaps close to the wall; again, she is shot into the stream, and perhaps is dragged over to the other side, where, caught in a whirlpool, she spins about. We can neither land nor run as we please. The boats are entirely unmanageable; no order in their running can be preserved; now one, now another, is ahead, each crew laboring for its own preservation.[7] In such a place we come to another rapid. Two of the boats run it perforce. One succeeds in landing, but there is no foothold by which to make a portage and she is pushed out again into the stream. The next minute a great reflex wave fills the open compartment; she is water-logged, and drifts unmanageable.[8] Breaker after breaker rolls over her and one capsizes her. The men are thrown out; but they cling to the boat, and she drifts down some distance alongside of us and we are able to catch her. She is soon bailed out and the men are aboard once more; but the oars are lost, and so a pair from the "Emma Dean" is spared. Then for two miles we find smooth water.

---

[6] Fend off, ward off.

[7] This problem is familiar to any canyon river-runner. The "lakes," boiling with eddies, that follow the great rapids can be as frustrating as the rapids are dangerous. It is in the lakes that the river must spend the enormous store of kinetic energy it has acquired from the fall of the rapids. It does this by creating a chaos of powerful backcurrents and whirlpools. Oars are frequently useless against the former, and some of the latter can swamp a boat.

[8] A reflex wave breaks upstream, and a boat that fails to punch through such a wave will fill quickly. Powell and his men probably were recovering from the eddies after Zoroaster Rapid when they encountered 85 Mile Rapid and capsized either the *Maid* or the *Sister,* as recounted in this passage.

Clouds are playing in the canyon today. Sometimes they roll down in great masses, filling the gorge with gloom; sometimes they hang aloft from wall to wall and cover the canyon with a roof of impending storm, and we can peer long distances up and down this canyon corridor, with its cloud-roof overhead, its walls of black granite, and its river bright with the sheen of broken waters. Then a gust of wind sweeps down a side gulch and, making a rift in the clouds, reveals the blue heavens, and a stream of sunlight pours in. Then the clouds drift away into the distance and hang around crags and peaks and pinnacles and towers and walls, and cover them with a mantle that lifts from time to time and sets them all in sharp relief. Then baby clouds creep out of side canyons, glide around points, and creep back again into more distant gorges. Then clouds arrange in strata across the canyon, with intervening vista views to cliffs and rocks beyond. The clouds are children of the heavens, and when they play among the rocks they lift them to the region above.

It rains! Rapidly little rills are formed above, and these soon grow into brooks, and the brooks grow into creeks and tumble over the walls in innumerable cascades, adding their wild music to the roar of the river. When the rain ceases the rills, brooks, and creeks run dry. The waters that fall during a rain on these steep rocks are gathered at once into the river; they could scarcely be poured in more suddenly if some vast spout ran from the clouds to the stream itself. When a storm bursts over the canyon a side gulch is dangerous, for a sudden flood may come, and the inpouring waters will raise the river so as to hide the rocks.

Early in the afternoon we discover a stream entering from the north—a clear, beautiful creek, coming down through a gorgeous red canyon. We land and camp on a sand beach above its mouth, under a great, overspreading tree with willow-shaped leaves.

August 16. We must dry our rations again to-day and make oars.

The Colorado is never a clear stream, but for the past three or four days it has been raining much of the time, and the floods poured over the walls have brought down great quantities of mud, making it exceedingly turbid now. The little affluent which we have discovered here is a clear, beautiful creek, or river, as it would be termed in this western country, where streams are not abundant. We have named one stream, away above, in honor of the great chief of the "Bad Angels," and as this is in beautiful contrast to that, we conclude to name it "Bright Angel."9

Early in the morning the whole party starts up to explore the Bright Angel River, with the special purpose of seeking timber from which to make oars. A couple of miles above we find a large pine log, which has been floated down from the plateau, probably from an altitude of more than 6,000 feet, but not many miles back. On its way it must have passed over many cataracts and falls, for it bears scars in evidence of the rough usage which it has received. The men roll it on skids, and the work of sawing oars is commenced.

This stream heads away back under a line of abrupt cliffs that terminates the plateau, and tumbles down more than 4,000 feet in the first mile or two of its course; then runs through a deep, narrow canyon until it reaches the river.

Late in the afternoon I return and go up a little gulch just above this creek, about 200 yards from camp, and discover the ruins of two or three old houses, which were originally of stone laid in mortar. Only the foundations are left, but irregular blocks, of which the houses were constructed, lie scattered about. In one room I find an old mealing-stone, deeply worn, as if it had been much used. A great deal of pottery is strewn around, and old trails, which in some places are deeply worn into the rocks, are seen.

It is ever a source of wonder to us why these ancient people sought

---

9 Bright Angel Creek enters the Colorado in the heart of the inner gorge. Sumner's account for this day calls the gorge "the worst hole in America, if not in the world." The tens of thousands of awestruck annual visitors to Grand Canyon National Park who reach the canyon bottom near here are under infinitely less duress and depart with much fonder feelings for the place. Phantom Ranch, a destination for many of those visitors, where limited supplies and services are available, lies a short distance up the Bright Angel Creek from the river. The creek Powell named for the chief of the "Bad Angels" is the Dirty Devil, which drains a large portion of central Utah. The expedition reached the Dirty Devil in Cataract Canyon on July 28, nearly 3 weeks earlier.

such inaccessible places for their homes. They were, doubtless, an agricultural race, but there are no lands here of any considerable extent that
they could have cultivated. To the west of Oraibi, one of the towns in the
Province of Tusayan,[10] in northern Arizona, the inhabitants have actually
built little terraces along the face of the cliff where a spring gushes out,
and thus made their sites for gardens. It is possible that the ancient inhabitants of this place made their agricultural lands in the same way. But why
should they seek such spots? Surely the country was not so crowded with
people as to demand the utilization of so barren a region. The only solution suggested of the problem is this: We know that for a century or two
after the settlement of Mexico many expeditions were sent into the country now comprising Arizona and New Mexico, for the purpose of bringing the town-building people under the dominion of the Spanish government. Many of their villages were destroyed, and the inhabitants fled to
regions at that time unknown; and there are traditions among the people
who inhabit the pueblos that still remain that the canyons were these
unknown lands. It may be these buildings were erected at that time; sure
it is that they have a much more modern appearance than the ruins scattered over Nevada, Utah, Colorado, Arizona, and New Mexico. Those
old Spanish conquerors had a monstrous greed for gold and a wonderful
lust for saving souls. Treasures they must have, if not on earth, why, then,
in heaven; and when they failed to find heathen temples bedecked with
silver, they propitiated Heaven by seizing the heathen themselves. There
is yet extant a copy of a record made by a heathen artist to express his
conception of the demands of the conquerors. In one part of the picture
we have a lake, and near by stands a priest pouring water on the head of a
native. On the other side, a poor Indian has a cord about his throat. Lines
run from these two groups to a central figure, a man with beard and full
Spanish panoply. The interpretation of the picture-writing is this: "Be
baptized as this saved heathen, or be hanged as that damned heathen."

---

[10] The land of the Hopi. Oraibi is a Hopi village.

Doubtless, some of these people preferred another alternative, and rather than be baptized or hanged they chose to imprison themselves within these canyon walls.[11]

August 17. Our rations are still spoiling; the bacon is so badly injured that we are compelled to throw it away. By an accident, this morning, the saleratus[12] was lost overboard. We have now only musty flour sufficient for ten days and a few dried apples, but plenty of coffee. We must make all haste possible. If we meet with difficulties such as we have encountered in the canyon above, we may be compelled to give up the expedition and try to reach the Mormon settlements to the north.

Our hopes are that the worst places are passed, but our barometers are all so much injured as to be useless and so we have lost our reckoning in altitude, and know not how much descent the river has yet to make.

The stream is still wild and rapid and rolls through a narrow channel. We make but slow progress, often landing against a wall and climbing around some point to see the river below. Although very anxious to advance, we are determined to run with great caution, lest by another accident we lose our remaining supplies. How precious that little flour has become! We divide it among the boats and carefully store it away, so that it can be lost only by the loss of the boat itself.

We make ten miles and a half, and camp among the rocks on the right. We have had rain from time to time all day, and have been thoroughly drenched and chilled; but between showers the sun shines with great power and the mercury in our thermometers stands at 115 degrees, so that

---

[11] Powell's characterization of Spanish settlement is typical for Anglo-Americans of his time, and, to be sure, Spanish colonization was harsh. The occupation of the Bright Angel site discovered by Powell, however, long predated the arrival of Spaniards and was probably complete by about 1140 (Stevens, *The Colorado River in Grand Canyon,* 20). Powell's assumption that "the country was not so crowded with people as to demand the utilization of so barren a region" probably is not accurate. Indeed, population density was low in relation to the vastness of the region, but it was high in relation to the availability of arable land. Few niches capable of supporting human life were left unoccupied during the period when the Bright Angel site was occupied.

[12] Baking soda.

we have rapid changes from great extremes, which are very disagreeable. It is especially cold in the rain to-night. The little canvas we have is rotten and useless; the rubber *ponchos* with which we started from Green River City have all been lost; more than half the party are without hats, not one of us has an entire suit of clothes, and we have not a blanket apiece. So we gather driftwood and build a fire; but after supper the rain, coming down in torrents, extinguishes it, and we sit up all night on the rocks, shivering, and are more exhausted by the night's discomfort than by the day's toil.

August 18. The day is employed in making portages[13] and we advance but two miles on our journey. Still it rains.

While the men are at work making portages I climb up the granite to its summit and go away back over the rust-colored sandstones and greenish-yellow shales to the foot of the marble wall. I climb so high that the men and boats are lost in the black depths below and the dashing river is a rippling brook, and still there is more canyon above than below. All about me are interesting geologic records. The book is open and I can read as I run. All about me are grand views, too, for the clouds are playing again in the gorges. But somehow I think of the nine days' rations and the bad river, and the lesson of the rocks and the glory of the scene are but half conceived.

I push on to an angle, where I hope to get a view of the country beyond, to see if possible what the prospect may be of our soon running through this plateau, or at least of meeting with some geologic change that will let us out of the granite; but, arriving at the point, I can see below only a labyrinth of black gorges.

August 19. Rain again this morning. We are in our granite prison still, and the time until noon is occupied in making a long, bad portage.

After dinner, in running a rapid the pioneer boat is upset by a wave. We are some distance in advance of the larger boats. The river is rough and swift and we are unable to land, but cling to the boat and are carried down

---

[13]Possibly around Granite and Hermit Rapids, two of the canyon's most difficult. Powell's estimate of making 10.5 miles the previous day, like most of his estimates of distance, was probably too great, especially because 2 miles below Bright Angel Creek the expedition would have had to contend with Horn Creek Rapid, another of the canyon's major trials.

stream over another rapid. The men in the boats above see our trouble, but they are caught in whirlpools and are spinning about in eddies, and it seems a long time before they come to our relief. At last they do come; our boat is turned right side up and bailed out; the oars, which fortunately have floated along in company with us, are gathered up, and on we go, without even landing. The clouds break away and we have sunshine again.

Soon we find a little beach with just room enough to land. Here we camp, but there is no wood. Across the river and a little way above, we see some driftwood lodged in the rocks. So we bring two boat loads over, build a huge fire, and spread everything to dry. It is the first cheerful night we have had for a week—a warm, drying fire in the midst of the camp, and a few bright stars in our patch of heavens overhead.

August 20. The characteristics of the canyon change this morning. The river is broader, the walls more sloping, and composed of black slates that stand on edge. These nearly vertical slates are washed out in places—that is, the softer beds are washed out between the harder, which are left standing. In this way curious little alcoves are formed, in which are quiet bays of water, but on a much smaller scale than the great bays and buttresses of Marble Canyon.

The river is still rapid and we stop to let down with lines several times, but make greater progress, as we run ten miles. We camp on the right bank. Here, on a terrace of trap, we discover another group of ruins. There was evidently quite a village on this rock. Again we find mealing-stones and much broken pottery, and up on a little natural shelf in the rock back of the ruins we find a globular basket that would hold perhaps a third of a bushel. It is badly broken, and as I attempt to take it up it falls to pieces. There are many beautiful flint chips, also, as if this had been the home of an old arrow-maker.

August 21. We start early this morning, cheered by the prospect of a fine day and encouraged also by the good run made yesterday. A quarter of a mile below camp the river turns abruptly to the left, and between camp and that point is very swift, running down in a long, broken chute and piling up against the foot of the cliff, where it turns to the left. We try to pull across, so as to go down on the other side, but the waters are swift and it seems impossible for us to escape the rock below; but, in pulling across, the bow of the boat is turned to the farther shore, so that we are swept broadside down and are prevented by the rebounding waters from striking against the wall. We toss about for a few seconds in these billows and are then carried past the danger. Below, the river turns again to the right, the canyon is very narrow, and we see in advance but a short distance. The water, too, is very swift, and there is no landing-place. From around this curve there comes a mad roar, and down we are carried with a dizzying velocity to the head of another rapid. On either side high over our heads there are overhanging granite walls, and the sharp bends cut off our view, so that a few minutes will carry us into unknown waters. Away we go on one long, winding chute. I stand on deck, supporting myself with a strap fastened on either side of the gunwale. The boat glides rapidly where the water is smooth, then, striking a wave, she leaps and bounds like a thing of life, and we have a wild, exhilarating ride for ten miles, which we make in less than an hour.[14] The excitement is so great that we forget the danger until we hear the roar of a great fall below; then we back on our oars and are carried slowly toward its head and succeed in landing just above and find that we have to make another portage. At this we are engaged until some time after dinner.

Just here we run out of the granite. Ten miles in less than half a day, and limestone walls below. Good cheer returns; we forget the storms and the gloom and the cloud-covered canyons and the black granite and the raging river, and push our boats from shore in great glee.

Though we are out of the granite, the river is still swift, and we wheel about a point again to the right, and turn, so as to head back in the direction from which we came; this brings the granite in sight again, with its narrow gorge and black crags; but we meet with no more great falls or rapids. Still, we run cautiously and stop from time to time to examine some

---

[14] One reason Powell consistently overstates the distances traveled by the expedition is that he overestimates the speed of the river. His maximum sustained rate of travel probably was no more than 8 miles per hour, rather than the 10 claimed here.

places which look bad. Yet we make ten miles this afternoon; twenty miles in all to-day.

August 22. We come to rapids again this morning and are occupied several hours in passing them, letting the boats down from rock to rock with lines for nearly half a mile, and then have to make a long portage.[15] While the men are engaged in this I climb the wall on the northeast to a height of about 2,500 feet, where I can obtain a good view of a long stretch of canyon below. Its course is to the southwest. The walls seem to rise very abruptly for 2,500 or 3,000 feet, and then there is a gently sloping terrace on each side for two or three miles, when we again find cliffs, 1,500 or 2,000 feet high. From the brink of these the plateau stretches back to the north and south for a long distance. Away down the canyon on the right wall I can see a group of mountains, some of which appear to stand on the brink of the canyon. The effect of the terrace is to give the appearance of a narrow winding valley with high walls on either side and a deep, dark, meandering gorge down its middle. It is impossible from this point of view to determine whether or not we have granite at the bottom; but from geologic considerations, I conclude that we shall have marble walls below.[16]

After my return to the boats we run another mile and camp for the night. We have made but little over seven miles to-day, and a part of our flour has been soaked in the river again.

August 23. Our way to-day is again through marble walls. Now and then we pass for a short distance through patches of granite, like hills thrust up into the limestone. At one of these places we have to make another portage, and, taking advantage of the delay, I go up a little stream to the north,[17] wading it all the way, sometimes having to plunge in to my neck, in other places being compelled to swim across little basins that have been excavated at the

---

[15] Deubendorf Rapid.

[16] The difference between the rocks was important to Powell for more than geological reasons. Granite, being more resistant than limestone, promised a narrower canyon, more terrible rapids, and fewer options for dealing with them.

[17] Tapeats Creek. A little over a mile up the creek, a side-stream joins from the west. Its entire flow is provided by Thunder Spring, which gushes loudly in clouds of spray from a vertical section of canyon wall.

foot of the falls. Along its course are many cascades and springs, gushing out from the rocks on either side. Sometimes a cottonwood tree grows over the water. I come to one beautiful fall, of more than 150 feet, and climb around it to the right on the broken rocks. Still going up, the canyon is found to narrow very much, being but 15 or 20 feet wide; yet the walls rise on either side many hundreds of feet, perhaps thousands; I can hardly tell.

In some places the stream has not excavated its channel down vertically through the rocks, but has cut obliquely, so that one wall overhangs the other. In other places it is cut vertically above and obliquely below, or obliquely above and vertically below, so that it is impossible to see out overhead. But I can go no farther; the time which I estimated it would take to make the portage has almost expired, and I start back on a round trot, wading in the creek where I must and plunging through basins. The men are waiting for me, and away we go on the river.

Just after dinner we pass a stream on the right, which leaps into the Colorado by a direct fall of more than 100 feet, forming a beautiful cascade.[18] There is a bed of very hard rock above, 30 or 40 feet in thickness, and there are much softer beds below. The hard beds above project many yards beyond the softer, which are washed out, forming a deep cave behind the fall, and the stream pours through a narrow crevice above into a deep pool below. Around on the rocks in the cavelike chamber are set beautiful ferns, with delicate fronds and enameled stalks. The frondlets have their points turned down to form spore cases. It has very much the appearance of the maidenhair fern, but is much larger. This delicate foliage covers the rocks all about the fountain, and gives the chamber great beauty. But we have little time to spend in admiration; so on we go.

We make fine progress this afternoon, carried along by a swift river, shooting over the rapids and finding no serious obstructions. The canyon walls for 2,500 or 3,000 feet are very regular, rising almost perpendicularly, but here and there set with narrow steps, and occasionally we can see, away above the broad terrace to distant cliffs.

---

[18]Deer Creek enters the Colorado via a spectacular falls.

We camp to-night in a marble cave, and find on looking at our reckoning that we have run 22 miles.

August 24. The canyon is wider to-day. The walls rise to a vertical height of nearly 3,000 feet. In many places the river runs under a cliff in great curves, forming amphitheaters half-dome shaped.

Though the river is rapid, we meet with no serious obstructions and run 20 miles. How anxious we are to make up our reckoning every time we stop, now that our diet is confined to plenty of coffee, a very little spoiled flour, and very few dried apples! It has come to be a race for a dinner. Still, we make such fine progress that all hands are in good cheer, but not a moment of daylight is lost.

August 25. We make 12 miles this morning, when we come to monuments of lava standing in the river, low rocks mostly, but some of them shafts more than a hundred feet high.[19] Going on down three or four miles, we find them increasing in number. Great quantities of cooled lava and many cinder cones are seen on either side; and then we come to an abrupt cataract.[20] Just over the fall on the right wall a cinder cone, or extinct volcano, with a well-defined crater, stands on the very brink of the canyon. This, doubtless, is the one we saw two or three days ago. From this volcano vast floods of lava have been poured down into the river, and a stream of molten rock has run up the canyon three or four miles and down we know not how far. Just where it poured over the canyon wall is the fall. The whole north side as far as we can see is lined with the black basalt, and high up on the opposite wall are patches of the same material, resting on the benches and filling old alcoves and eaves, giving the wall a spotted appearance.

The rocks are broken in two along a line which here crosses the river, and the beds we have seen while coming down the canyon for the last 30 miles have dropped 800 feet on the lower side of the line, forming what geologists call a "fault." The volcanic cone stands directly over the fissure thus formed. On the left side of the river, opposite, mammoth springs burst

---

[19]Vulcan's Anvil, a mile above Lava Falls, is the most impressive of these.
[20]Lava Falls, which is universally acknowledged to be the most awesome stretch of whitewater in the canyon.

out of this crevice, 100 or 200 feet above the river, pouring in a stream quite equal in volume to the Colorado Chiquito.[21]

This stream seems to be loaded with carbonate of lime, and the water, evaporating, leaves an incrustation on the rocks; and this process has been continued for a long time, for extensive deposits are noticed in which are basins with bubbling springs. The water is salty.

We have to make a portage here, which is completed in about three hours; then on we go.

We have no difficulty as we float along, and I am able to observe the wonderful phenomena connected with this flood of lava. The canyon was doubtless filled to a height of 1,200 or 1,500 feet, perhaps by more than one flood. This would dam the water back; and in cutting through this great lava bed, a new channel has been formed, sometimes on one side, sometimes on the other.[22] The cooled lava, being of firmer texture than the rocks of which the walls are composed, remains in some places; in others a narrow channel has been cut, leaving a line of basalt on either side. It is possible that the lava cooled faster on the sides against the walls and that the center ran out; but of this we can only conjecture. There are other places where almost the whole of the lava is gone, only patches of it being seen where it has caught on the walls. As we float down we can see that it ran out into side canyons. In some places this basalt has a fine, columnar structure, often in concentric prisms, and masses of these concentric columns have coalesced. In some places, when the flow occurred the canyon was probably about the same depth that it is now, for we can see where the basalt has rolled out on the sands, and—what seems curious to me—the sands are not melted or metamorphosed to any appreciable extent. In places the bed of the river is of sandstone or limestone, in other places of lava, showing that it has all been cut out again where the sandstones and limestones appear; but there is a little yet left where the bed is of lava.

What a conflict of water and fire there must have been here! Just imagine

---

[21]The Little Colorado River.
[22]In places along this stretch of river one can see former river channels perched on benches high above the present level of the river.

a river of molten rock running down into a river of melted snow. What a seething and boiling of the waters; what clouds of steam rolled into the heavens!

Thirty-five miles to-day. Hurrah!

August 26. The canyon walls are steadily becoming higher as we advance. They are still bold and nearly vertical—up to the terrace. We still see evidence of the eruption discovered yesterday, but the thickness of the basalt is decreasing as we go down stream; yet it has been reinforced at points by streams that have come down from volcanoes standing on the terrace above, but which we cannot see from the river below.

Since we left the Colorado Chiquito we have seen no evidences that the tribe of Indians inhabiting the plateaus on either side ever come down to the river; but about eleven o'clock to-day we discover an Indian garden at the foot of the wall on the right, just where a little stream with a narrow flood plain comes down through a side canyon. Along the valley the Indians have planted corn, using for irrigation the water which bursts out in springs at the foot of the cliff. The corn is looking quite well, but it is not sufficiently advanced to give us roasting ears; but there are some nice green squashes. We carry ten or a dozen of these on board our boats and hurriedly leave, not willing to be caught in the robbery, yet excusing ourselves by pleading our great want. We run down a short distance to where we feel certain no Indian can follow, and what a kettle of squash sauce we make! True, we have no salt with which to season it, but it makes a fine addition to our unleavened bread and coffee. Never was fruit so sweet as these stolen squashes.

After dinner we push on again and make fine time, finding many rapids, but none so bad that we cannot run them with safety; and when we stop, just at dusk, and foot up our reckoning, we find we have run 35 miles again. A few days like this, and we are out of prison.[23]

---

[23] Powell knows that the Virgin River joins the Colorado in open country, which will provide release from the "prison" of the canyon. He also knows the difference in longitude between the mouth of the Virgin and other points of reference earlier in the journey, from which he can calculate the total mileage of westward progress he must make. At this point, he does not know precisely how much farther off the end of his journey lies, but he knows he is getting close.

We have a royal supper—unleavened bread, green squash sauce, and strong coffee. We have been for a few days on half rations, but now have no stint of roast squash.

August 27. This morning the river takes a more southerly direction. The dip of the rocks is to the north and we are running rapidly into lower formations. Unless our course changes we shall very soon run again into the granite. This gives some anxiety. Now and then the river turns to the west and excites hopes that are soon destroyed by another turn to the south. About nine o'clock we come to the dreaded rock. It is with no little misgiving that we see the river enter these black, hard walls. At its very entrance we have to make a portage; then let down with lines past some ugly rocks. We run a mile or two farther, and then the rapids below can be seen.

About eleven o'clock we come to a place in the river which seems much worse than any we have yet met in all its course.[24] A little creek comes down from the left. We land first on the right and clamber up over the granite pinnacles for a mile or two, but can see no way by which to let down, and to run it would be sure destruction. After dinner we cross to examine on the left. High above the river we can walk along on the top of the granite, which is broken off at the edge and set with crags and pinnacles, so that it is very difficult to get a view of the river at all. In my eagerness to reach a point where I can see the roaring fall below, I go too far on the wall, and can neither advance nor retreat. I stand with one foot on a little projecting rock and cling with my hand fixed in a little crevice. Finding I am caught here, suspended 400 feet above the river, into which I must fall if my footing fails, I call for help. The men come and pass me a line, but I cannot let go of the rock long enough to take hold of it. Then they bring two or three of the largest oars. All this takes time which seems very precious to me; but at last they arrive. The blade of one of the oars is pushed into a little crevice in the rock beyond me in such a manner that they can hold me pressed against the

---

[24] This stretch of whitewater will bear the name Separation Rapid, for reasons that will soon be evident. Today, however, the rapid is drowned, absorbed in the slackening flow of the river as it enters Lake Mead.

wall. Then another is fixed in such a way that I can step on it; and thus I am extricated.

Still another hour is spent in examining the river from this side, but no good view of it is obtained; so now we return to the side that was first examined, and the afternoon is spent in clambering among the crags and pinnacles and carefully scanning the river again. We find that the lateral streams have washed boulders into the river, so as to form a dam, over which the water makes a broken fall of 18 or 20 feet; then there is a rapid, beset with rocks, for 200 or 300 yards, while on the other side, points of the wall project into the river. Below, there is a second fall how great, we cannot tell. Then there is a rapid; filled with huge rocks, for 100 or 200 yards. At the bottom of it, from the right wall, a great rock projects quite halfway across the river. It has a sloping surface extending up stream, and the water, coming down with all the momentum gained in the falls and rapids above, rolls up this inclined plane many feet, and tumbles over to the left. I decide that it is possible to let down over the first fall, then run near the right cliff to a point just above the second, where we can pull out into a little chute, and, having run over that in safety, if we pull with all our power across the stream, we may avoid the great rock below. On my return to the boat I announce to the men that we are to run it in the morning. Then we cross the river and go into camp for the night on some rocks in the mouth of the little side canyon.

After supper Captain Howland asks to have a talk with me. We walk up the little creek a short distance, and I soon find that his object is to remonstrate against my determination—to proceed. He thinks that we had better abandon the river here. Talking with him, I learn that he, his brother, and William Dunn have determined to go no farther in the boats. So we return to camp. Nothing is said to the other men.

For the last two days our course has not been plotted. I sit down and do this now, for the purpose of finding where we are by dead reckoning. It is a clear night, and I take out the sextant to make observation for latitude, and I find that the astronomic determination agrees very nearly with that of the plot—quite as closely as might be expected from a meridian observation on a planet. In a direct line, we must be about 45 miles from the mouth of the Rio Virgen.[25] If we can reach that point, we know that there are settlements up that river about 20 miles. This 45 miles in a direct line will probably be 80 or 90 by the meandering line of the river. But then we know that there is comparatively open country for many miles above the mouth of the Virgen, which is our point of destination.

As soon as I determine all this, I spread my plot on the sand and wake Howland, who is sleeping down by the river, and show him where I suppose we are, and where several Mormon settlements are situated.

We have another short talk about the morrow, and he lies down again; but for me there is no sleep. All night long I pace up and down a little path, on a few yards of sand beach, along by the river. Is it wise to go on? I go to the boats again to look at our rations. I feel satisfied that we can get over the danger immediately before us; what there may be below I know not. From our outlook yesterday on the cliffs, the canyon seemed to make another great bend to the south, and this, from our experience heretofore, means more and higher granite walls. I am not sure that we can climb out of the canyon here, and, if at the top of the wall, I know enough of the country to be certain that it is a desert of rock and sand between this and the nearest Mormon town, which, on the most direct line, must be 75 miles away. True, the late rains have been favorable to us, should we go out, for the probabilities are that we shall find water still standing in holes; and at one time I almost conclude to leave the river. But for years I have been contemplating this trip. To leave the exploration unfinished, to say that there is a part of the canyon which I cannot explore, having already nearly accomplished it, is more than I am willing to acknowledge, and I determine to go on.

I wake my brother and tell him of Howland's determination, and he promises to stay with me; then I call up Hawkins, the cook, and he makes a like promise; then Sumner and Bradley and Hall, and they all agree to go on.

August 28. At last daylight comes and we have breakfast without a word being said about the future. The meal is as solemn as a funeral. After break-

---

[25] The Virgin River of Utah and Nevada.

fast I ask the three men if they still think it best to leave us. The elder Howland thinks it is, and Dunn agrees with him. The younger Howland tries to persuade them to go on with the party; failing in which, he decides to go with his brother.

Then we cross the river. The small boat is very much disabled and unseaworthy. With the loss of hands, consequent on the departure of the three men, we shall not be able to run all of the boats; so I decide to leave my "Emma Dean."

Two rifles and a shotgun are given to the men who are going out. I ask them to help themselves to the rations and take what they think to be a fair share. This they refuse to do, saying they have no fear but that they can get something to eat; but Billy, the cook, has a pan of biscuits prepared for dinner, and these he leaves on a rock.

Before starting, we take from the boat our barometers, fossils, the minerals, and some ammunition and leave them on the rocks. We are going over this place as light as possible. The three men help us lift our boats over a rock 25 or 30 feet high and let them down again over the first fall, and now we are all ready to start. The last thing before leaving, I write a letter to my wife and give it to Howland. Sumner gives him his watch, directing that it be sent to his sister should he not be heard from again. The records of the expedition have been kept in duplicate. One set of these is given to Howland;[26] and now we are ready. For the last time they entreat us not to go on, and tell us that it is madness to set out in this place; that we can never get safely through it; and, further, that the river turns again to the south into the granite, and a few miles of such rapids and falls will exhaust our entire stock of rations, and then it will be too late to climb out. Some tears are shed; it is rather a solemn parting; each party thinks the other is taking the dangerous course.

My old boat left, I go on board the "Maid of the Canyon." The three men

---

[26] Powell mistakenly gives Howland both sets of notes from the first phase of the trip, an oversight that later helps justify a second exploration of the Colorado when both sets are lost with Howland's death.

climb a crag that overhangs the river to watch us off. The "Maid of the Canyon" pushes out. We glide rapidly along the foot of the wall, just grazing one great rock, then pull out a little into the chute of the second fall and plunge over it. The open compartment is filled when we strike the first wave below, but we cut through it, and then the men pull with all their power toward the left wall and swing clear of the dangerous rock below all right. We are scarcely a minute in running it, and find that, although it looked bad from above, we have passed many places that were worse.

The other boat follows without more difficulty. We land at the first practicable point below, and fire our guns, as a signal to the men above that we have come over in safety. Here we remain a couple of hours, hoping that they will take the smaller boat and follow us. We are behind a curve in the canyon and cannot see up to where we left them, and so we wait until their coming seems hopeless, and then push on.

And now we have a succession of rapids and falls until noon, all of which we run in safety. Just after dinner we come to another bad place. A little stream comes in from the left, and below there is a fall, and still below another fall.[27] Above, the river tumbles down, over and among the rocks, in whirlpools and great waves, and the waters are lashed into mad, white foam. We run along the left, above this, and soon see that we cannot get down on this side, but it seems possible to let down on the other. We pull up stream again for 200 or 300 yards and cross. Now there is a bed of basalt on this northern side of the canyon, with a bold escarpment that seems to be a hundred feet high. We can climb it and walk along its summit to a point where we are just at the head of the fall. Here the basalt is broken down again, so it seems to us, and I direct the men to take a line to the top of the cliff and let the boats down along the wall. One man remains in the boat to keep her clear of the rocks and prevent her line from being caught on the projecting angles. I climb the cliff and pass along to a point just over the fall and descend by broken rocks, and find that the break of the fall is above the break of the wall, so that we cannot land, and that still below the river is very

---

[27] This was Lava Cliff Rapid; Lake Mead has since drowned it.

bad, and that there is no possibility of a portage. Without waiting further to examine and determine what shall be done, I hasten back to the top of the cliff to stop the boats from coming down. When I arrive I find the men have let one of them down to the head of the fall. She is in swift water and they are not able to pull her back; nor are they able to go on with the line, as it is not long enough to reach the higher part of the cliff which is just before them; so they take a bight around a crag. I send two men back for the other line. The boat is in very swift water, and Bradley is standing in the open compartment, holding out his oar to prevent her from striking against the foot of the cliff. Now she shoots out into the stream and up as far as the line will permit, and then, wheeling, drives headlong against the rock, and then out and back again, now straining on the line, now striking against the rock. As soon as the second line is brought, we pass it down to him; but his attention is all taken up with his own situation, and he does not see that we are passing him the line. I stand on a projecting rock, waving my hat to gain his attention, for my voice is drowned by the roaring of the falls. Just at this moment I see him take his knife from its sheath and step forward to cut the line.

He has evidently decided that it is better to go over with the boat as it is than to wait for her to be broken to pieces. As he leans over, the boat sheers again into the stream, the stem-post breaks away and she is loose. With perfect composure Bradley seizes the great scull oar, places it in the stern rowlock, and pulls with all his power (and he is an athlete) to turn the bow of the boat down stream, for he wishes to go bow down, rather than to drift broad side on. One, two strokes he makes, and a third just as she goes over, and the boat is fairly turned, and she goes down almost beyond our sight, though we are more than a hundred feet above the river. Then she comes up again on a great wave, and down and up, then around behind some great rocks, and is lost in the mad, white foam below. We stand frozen with fear, for we see no boat. Bradley is gone! so it seems. But now, away below, we see something coming out of the waves. It is evidently a boat. A moment more, and we see Bradley standing on deck, swinging his hat to show that he is all right. But he is in a whirlpool. We have the stem-post of his boat attached to the line. How badly she may be disabled we know not. I direct Sumner and Powell to pass along the cliff and see if they can reach him from below. Hawkins, Hall, and myself run to the other boat, jump aboard, push out, and away we go over the falls. A wave rolls over us and our boat is unmanageable. Another great wave strikes us, and the boat rolls over, and tumbles and tosses, I know not how. All I know is that Bradley is picking us up. We soon have all right again, and row to the cliff and wait until Sumner and Powell can come. After a difficult climb they reach us. We run two or three miles farther and turn again to the northwest, continuing until night, when we have run out of the granite once more.

August 29. We start very early this morning. The river still continues swift, but we have no serious difficulty, and at twelve o'clock emerge from the Grand Canyon of the Colorado. We are in a valley now, and low mountains are seen in the distance, coming to the river below. We recognize this as the Grand Wash.

A few years ago a party of Mormons set out from St. George, Utah, taking with them a boat, and came down to the Grand Wash, where they divided, a portion of the party crossing the river to explore the San Francisco Mountains. Three men—Hamblin,[28] Miller, and Crosby—taking the boat, went on down the river to Callville, landing a few miles below the mouth of the Rio Virgen. We have their manuscript journal with us, and so the stream is comparatively well known.

To-night we camp on the left bank, in a mesquite thicket.

The relief from danger and the joy of success are great. When he who has been chained by wounds to a hospital cot until his canvas tent seems like a dungeon cell, until the groans of those who lie about tortured with probe and knife are piled up, a weight of horror on his ears that he cannot throw off, cannot forget, and until the stench of festering wounds and anaesthetic drugs has filled the air with its loathsome burthen,—when he at

---

[28] Jacob Hamblin was a missionary, pathfinder, and scout for Brigham Young's expanding Mormon colony of Deseret. Hamblin would accompany Powell on several important investigations of the canyon country in the years ahead.

last goes out into the open field, what a world he sees! How beautiful the sky, how bright the sunshine, what "floods of delirious music" pour from the throats of birds, how sweet the fragrance of earth and tree and blossom! The first hour of convalescent freedom seems rich recompense for all pain and gloom and terror.

Something like these are the feelings we experience to-night. Ever before us has been an unknown danger, heavier than immediate peril. Every waking hour passed in the Grand Canyon has been one of toil. We have watched with deep solicitude the steady disappearance of our scant supply of rations, and from time to time have seen the river snatch a portion of the little left, while we were a-hungered. And danger and toil were endured in those gloomy depths, where ofttimes clouds hid the sky by day and but a narrow zone of stars could be seen at night. Only during the few hours of deep sleep, consequent on hard labor, has the roar of the waters been hushed. Now the danger is over, now the toil has ceased, now the gloom has disappeared, now the firmament is bounded only by the horizon, and what a vast expanse of constellations can be seen!

The river rolls by us in silent majesty; the quiet of the camp is sweet; our joy is almost ecstasy. We sit till long after midnight talking of the Grand Canyon, talking of home, but talking chiefly of the three men who left us. Are they wandering in those depths, unable to find a way out? Are they searching over the desert lands above for water? Or are they nearing the settlements?

August 30. We run in two or three short, low canyons to-day, and on emerging from one we discover a band of Indians in the valley below. They see us, and scamper away in eager haste to hide among the rocks. Although we land and call for them to return, not an Indian can be seen.

Two or three miles farther down, in turning a short bend of the river, we come upon another camp. So near are we before they can see us that I can shout to them, and, being able to speak a little of their language, I tell them we are friends; but they all flee to the rocks, except a man, a woman, and two children. We land and talk with them. They are without lodges, but have built little shelters of boughs, under which they wallow in the sand. The man is dressed in a hat; the woman, in a string of beads only. At first they are evidently much terrified; but when I talk to them in their own language and tell them we are friends, and inquire after people in the Mormon towns, they are soon reassured and beg for tobacco. Of this precious article we have none to spare. Sumner looks around in the boat for something to give them, and finds a little piece of colored soap, which they receive as a valuable present, rather as a thing of beauty than as a useful commodity, however. They are either unwilling or unable to tell us anything about the Indians or white people, and so we push off, for we must lose no time.

We camp at noon under the right bank. And now as we push out we are in great expectancy, for we hope every minute to discover the mouth of the Rio Virgen. Soon one of the men exclaims: "Yonder's an Indian in the river." Looking for a few minutes, we certainly do see two or three persons. The men bend to their oars and pull toward them. Approaching, we see that there are three white men and an Indian hauling a seine, and then we discover that it is just at the mouth of the long-sought river.

As we come near, the men seem far less surprised to see us than we do to see them. They evidently know who we are, and on talking with them they tell us that we have been reported lost long ago, and that some weeks before a messenger had been sent from Salt Lake City with instructions for them to watch for any fragments or relics of our party that might drift down the stream.

Our new-found friends, Mr. Asa and his two sons, tell us that they are pioneers of a town that is to be built on the bank. Eighteen or twenty miles up the valley of the Rio Virgen there are two Mormon towns, St. Joseph and St. Thomas. To-night we dispatch an Indian to the last-mentioned place to bring any letters that may be there for us.

Our arrival here is very opportune. When we look over our store of supplies, we find about 10 pounds of flour, 15 pounds of dried apples, but 70 or 80 pounds of coffee.

August 31. This afternoon the Indian returns with a letter informing us that Bishop Leithhead of St. Thomas and two or three other Mormons are coming down with a wagon, bringing us supplies. They arrive about sundown. Mr. Asa treats us with great kindness to the extent of his ability; but Bishop Leithhead brings in his wagon two or three dozen melons, and many other little luxuries, and we are comfortable once more.

September 1. This morning Sumner, Bradley, Hawkins, and Hall, taking on a small supply of rations, start down the Colorado with the boats. It is their intention to go to Fort Mojave, and perhaps from there overland to Los Angeles.

Captain Powell and myself return with Bishop Leithhead to St. Thomas. From St. Thomas we go to Salt Lake City.

PART III

Among the Natives
of the Colorado Plateau

I N 1870 POWELL undertook a lengthy reconnaissance of the plateau country north of the Grand Canyon, much of the time in the company of Jacob Hamblin, a Mormon missionary and frontiersman, whom we will shortly meet. Powell sought not just to understand better the lay of the land but to extend and organize his insights into its geologic history—to understand how the land had attained its present form. Along the way he also intended to learn how to provide support and supplies for the second river expedition he was already planning. The knowledge of Indian guides would prove essential to the success of Powell's research, and he eagerly and repeatedly recruited assistance from a succession of them from the Southern Paiute bands he encountered. As he traveled the plateau country, Powell observed his guides and their families as closely as he observed the landscape. Among his Indian companions were a small band of Shivwits who admitted to having killed the two Howlands and William Dunn after they abandoned Powell's first river expedition the preceding year. Powell accepted their account, and probably so should we.

A contradictory explanation does exist for the three deaths, however. In recent years a letter surfaced that refers to "the day those three were murdered in our ward & the murderer killed to stop the shedding of more blood." William Leany, a Mormon pioneer, wrote the letter in 1883 to another pioneer of the Virgin River country, and in it he goes on a second time to mention, "the killing [of] the three in one room in our own ward." Although the letter gives no date and no further detail for the murders, it has fueled speculation that Mormons, not Indians, killed the three men from the Powell expedition.[1]

There was much dissension and factionalism in the Mormon communities of southern Utah in the 1860s and 1870s, and fear of "gentile" aggression fed animosity toward outsiders. No doubt some such outsiders and more than a few Mormon insiders were murdered in southern Utah in those days. Nevertheless, the argument that one or more whites killed the Howlands and Dunn founders for want of stronger evidence. It also rests on an assumption that Jacob Hamblin, who translated conversations between Powell and the Shivwits, was complicit in covering up Mormon involvement in the deed. According to this line of thinking, Hamblin distorted what the Indians said to convince Powell that the Shivwits were the culprits. Such a conspiracy theory further assumes not only that Hamblin was duplicitous but that Powell was easily duped. Neither idea fits well with what is known about the two men. Powell was in a position to observe closely the behavior of all parties, including the Shivwits, and he was then developing a working knowledge of their language. Moreover, he and Hamblin, his respected trailmate and companion, traveled together off and on many miles and months. For such a charade to have transpired, Powell would have had to misinterpret a great deal of what went on around him, and that seems unlikely.

Another point that bears remarking concerns Powell's depiction of the Southern Paiutes in the next selections. His patronizing language and the obsequiousness he attributes to the Indians would rightly be considered offensive coming from a white ethnographer or anthropologist today. By the standards of his day and culture, however, Powell showed marked sympathy in word and deed to Native Americans as individuals, as well as to their cultures and concerns. As is clear in this and other selections, his interest in Indian lifeways was genuine and respectful, a perspective that was by no means widely shared among whites. To place Powell in context, it is worth noting that when he wrote the following passages, Custer's catas-

---

[1]Larsen, "The 'Letter' or Were the Powell Men Really Killed by Indians," 14.

trophe at the Little Bighorn still lay more than a year in the future; conflict between whites and the Chiricahua, which stirred newspapers in Arizona and New Mexico to shriek for Apache blood, had more than a decade left to run; and the horrid Sioux tragedy and American national dishonor of Wounded Knee would not occur for 15 more years.

But if Powell's interest and compassion were exceptional for his day, he was in no way a cultural relativist. His belief in the superiority of his own cultural inheritance was adamant and unshakeable, and as he makes clear at the conclusion of the following selection, he believed there was no possibility that the Indians of the American West might successfully resist the "march of humanity" and remain in "their aboriginal condition." Instead, Powell held that the best hope for Indian survival was to be absorbed into white society, and throughout his career he urged that Indians adopt white values.

Putting aside the patronizing manner in which Powell describes the Southern Paiutes, one cannot quibble with the content of what he describes. The Southern Paiutes in general and the Shivwits in particular lived a harsh existence. Consider the report of Thomas J. Farham, who wrote in 1843 that Paiutes were "hunted in the spring of the year, when weak and helpless, by a certain class of men, and when taken, are fattened, carried to Santa Fe and sold as slaves during their minority." Although trade in human captives was long widespread throughout the West, the fate of being hunted when "weak and helpless" rarely befell the natives of richer lands, who could better resist the oppression of both winter and their adversaries.[2]

The second selection in this part, a magazine article titled "The Ancient Province of Tusayan," is one of three that Powell published in *Scribner's Monthly* in 1875, and it affords a marvelous vantage from which to appreciate Powell's sensibilities. His ethnographic eye is constantly at work, capturing details of the culture and behavior of all whom he encounters, Mormon pioneers no less than Indians, Paiutes no less than Hopis. At every turn he appraises the landscape, quickly drawing conclusions, usually correct ones, about how it was formed. He draws on seemingly boundless energy to scamper up the Vermilion Cliffs and gaze upon Marble Canyon

As quoted by Kelly and Fowler, "Southern Paiute," 386.

before catching up with the rest of his party half a day later, or to gallop off with Jacob Hamblin for 20 miles to investigate a volcanic dike that Mormon lore had made out to be the masonry of giants. One senses in these pages Powell's genuine delight in probing the differentness of other people and cultures. And certainly he was not immune to thrilling in the exoticness of what he encountered. Imagine him, as he would have us do, sitting naked for hours in a Hopi kiva while equally naked priests and bare-breasted virgins perform strange and solemn rituals, all the while struggling to remember the sequence and the meaning of the rites. *Scribner's* readers must have reveled in the unalloyed outlandishness of it all.

At every instant, Powell was pursuing research of the most serious nature. From his earliest contact with Native Americans (and especially through the winter of 1867–68 while camped on the White River), Powell had compiled Indian vocabularies, and he was now far along in building the family tree of North American Indian languages, one of his most enduring contributions to the anthropology of the continent. For instance, he had deduced, apparently on this very trip to Oraibi, that the language of the Hopi shared a common ancestor with the languages of the Paiutes and Utes. Powell was also involved in diplomacy, although one gets no hint of it from his *Scribner's* articles. After 2 months at Oraibi, Powell paused in his homebound journey to Santa Fe to stop with Hamblin at Fort Defiance, Arizona, for a peace conference with the Navajo. Although he lacked authority to represent the government in any way, Powell threw himself into the proceedings, making speeches to the Navajo on the necessity of the reservation system and arguing the benefits of peaceful relations between Navajos and Mormons. The result was an outcome long sought by Hamblin: a pledge by the Navajos to cease raiding Mormons across the Colorado River in Utah.[3]

---

[3]This was not the end of Powell's Indian diplomacy. In 1873 he accepted appointment as a special commissioner to the tribes of Utah and eastern Nevada for the purpose of obtaining an accurate census and persuading the tribes to accept reservation life. Characteristically, Powell turned his commission into something more, using it to campaign in Washington for a policy that treated reservations not merely as corrals for sequestering Indians away from whites but as places where Indians might receive instruction in skills that would enable them to support themselves.

As one historian has called him, Powell was "a scientist of brilliant lucidity with the imagination of a conjurer."[4] Popularization came as naturally to him as science, and it never threw him off the track of deeper investigations. As he told the leader of the Shivwits, he liked to travel with a photographer and "take pictures of everything." The resulting plates might lead to an engraving in *Scribner's* or an ethnographic portrait. Either one was as easy—and as hard—to produce. There was nothing that did not interest him and nothing that did not find a place in the mosaic of his larger understanding and sense of purpose. In the years after his first descent of the Colorado, this son of a Methodist missionary felt a mission taking shape within himself to comprehend the true character of the American West, its land, and its people and to teach his compatriots what that character portended for them and their civilization.

Goetzmann, *Exploration and Empire*, 566.

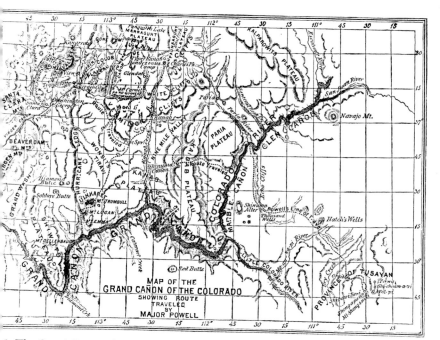

1. The Grand Cañon of Colorado, showing route traveled by Major Powell. From *Scribner's*
hly 10/6 (October 1875): 661. Courtesy Fondren Library, Southern Methodist University.

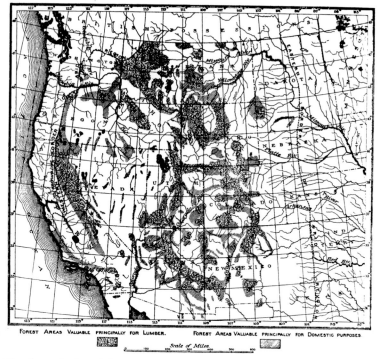

2. The forest lands of the arid region. From *Century Magazine* 40 (April 1890): 918.
sy Fondren Library, Southern Methodist University.

Map 3. The Utah Territory representing the extent of irrigable, timber, and pasture lands, From *Report on the Lands of the Arid Region,* 2nd ed., 1879. Courtesy DeGolyer I Southern Methodist University.

ARID REGION
OF THE
UNITED STATES
Showing Drainage Districts

Map 7. Arid region of the United States showing drainage districts, entire. From *Eleventh Annual Report of the United States Geological Survey*, 1890–91. Courtesy Dan Flores.

. Linguistic stocks of American Indians North of Mexico, detail. Courtesy DeGolyer Library,
ern Methodist University.

Map 5. Linguistic stocks of American Indians North of Mexico, entire. From *Seventh Annual Report of the Bureau of Ethnology*, 1891. Courtesy DeGolyer Library, Southern Methodist University.

4. The Utah Territory representing the extent of irrigable, timber, and pasture lands, detail show-
[s]alt Lake City, Utah Lake, and Southern Wasatch Mountains. "Irrigable Land" is light green,
[sha]ding Timber" is dark green, and "Area destitute of Timber on account of fires" is reddish tan.
[Cour]tesy DeGolyer Library, Southern Methodist University.

The following is excerpted from Powell's testimony to the House Select Committee on Irrigation, March 15, 1890, as reproduced in the Eleventh Annual Report of the United States Geological Survey, Part II: The Second Annual Report of the Irrigation Survey *(Washington: Government Printing Office, 1891), pp. 55–57.*

My theory is to organize in the United States another unit of government for specific purposes, for agriculture by irrigation, for the protection of the forests which are being destroyed by fire, and for the utilization of the pasturage which can only be utilized in large bodies; that is to create a great body of commonwealths. In the main these commonwealths would be like county communities in the States. . . .

Say to the States, If you will allow the people, wherever these interstate districts are found, to organize solely for the purpose of controlling the water, we will turn over all to them. [The General Government] will not give the lands, but [it] will declare that the pasturage lands and the timber lands are held by that General Government as the custodian of the people, and they are allowed the benefit and use of the timber and pasturage thereof, but that no individual shall get control of either the timber or pasturage lands. . . .

Let them make their own laws to govern the use of that timber in their own way and govern the pasturage in their own way. If they want the timber destroyed, if they want to sell it, if they want to destroy and wipe out irrigation, they are responsible for it, and let them do as they please. Say to them, You cannot sell this land, you need the wood, and you need this timber for your farms, and if you protect it from fires and cut it in such a manner that it will not injure your rivers and sources of supply for irrigation, you may have the timber. . . .

Let the General Government organize the arid region, including all of the lands to be irrigated by perennial streams, into irrigation districts by hydrographic basins. . . . Then let the people of each such irrigation district organize as a body and control the waters on the declared irrigable lands in any manner which they may devise. Then declare that the pasturage and timber lands be permanently reserved for the purpose for which they are adopted, and give to the people the right to protect and use the forests and the grasses. Let the Government retain the ownership of reservoir sites, canal sites, and headwork sites; but allow the people of each district to use them, as a body, so as to prevent speculation in such sites, which would ultimately be a tax on agriculture.

Some of these districts would lie in two States. To this arrangement the consent of the States should be obtained, and all the districts should be organized under State laws. The Government should not grant these privileges to the districts until the States themselves ratify the agreement and provide statutes for the organization of the districts and for the regulation of water rights, the protection and use of forests, and the protection and use of pasturage. This is the general plan which I present. There are minor questions to be considered, but the fundamental principles of the system are simple, as I have stated them.

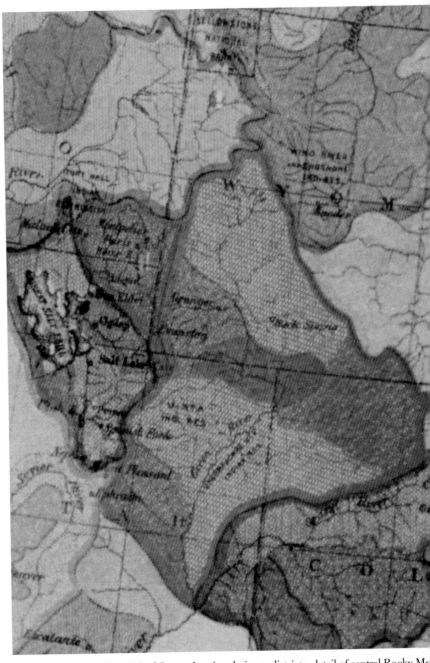

Map 8. Arid region of the United States showing drainage districts, detail of central Rocky M[...]
Courtesy Dan Flores.

—◇—

Camped with the Shivwits and
the Fate of the Separated Three

———

The year is 1870. *We join Powell, Hamblin, and their party, which includes a band of Kaibab Paiutes, encamped on the rim of the Grand Canyon.*

———

The next day, September 19, we were tired and sore, and concluded to rest a day with our Indian neighbors. During the inclement season they live in shelters made of boughs or the bark of the cedar, which they strip off in long shreds. In this climate, most of the year is dry and warm, and during such time they do not care for shelter. Clearing a small, circular space of ground, they bank it around with brush and sand, and wallow in it during the day,—and huddle together in a heap at night, men, women, and children; buckskin, rags, and sand. They wear very little clothing, not needing much in this lovely climate.

Altogether, these Indians are more nearly in their primitive condition than any others on the continent with whom I am acquainted. They have never received anything from the Government and are too poor to tempt the trader, and their country is so nearly inaccessible that the white man never visits them. The sunny mountain-side is covered with wild fruits, nuts, and native grains, upon which they subsist. The *oose*, the fruit of the yucca, or Spanish bayonet, is rich, and not unlike the pawpaw of the valley

This passage is excerpted from "An Overland Trip to the Grand Cañon," *Scribner's Monthly* 10/13 (October 1875): 659–78. Most of it also appears in slightly different form in *The Exploration of the Colorado River*, 316–23.

99

of the Ohio. They eat it raw and also roast it in the ashes. They gather the fruits of a cactus plant,[1] which are rich and luscious, and eat them as grapes or express the juice from them, making the dry pulp into cakes and saving them for winter; the wine they drink about their camp-fires until the midnight is merry with their revelries.

They also gather the seeds of many plants, as sunflowers, golden-rod, and grasses. For this purpose they have large conical baskets, which hold two or more bushels. The women carry them on their backs, suspended from their foreheads by broad straps, and with a smaller one in the left hand and a willow-woven fan in the right they walk among the grasses and sweep the seed into the smaller basket, which is emptied now and then into the larger, until it is full of seed and chaff. Then they winnow out the chaff and roast the seeds by a curious process: the seeds with a quantity of red-hot coals are put into a willow tray and, by rapidly and dexterously shaking and tossing them, they keep the coals aglow and the seeds and tray from burning. As if by magic, so skilled are the crones in this work they roll the seeds to one side of the tray as they are roasted and the coals to the other.

Then they grind the seeds into a fine flour and make it into cakes and mush. For a mill, they use a large flat rock, lying on the ground, and another small cylindrical one in the hands. They sit prone on the ground, holding the large flat rock between the feet and legs, then fill their laps with seeds, thus making a hopper to the mill with their dusky legs, and grind by pushing the seeds across the larger rock, where it drops into a tray. It is a merry sight to see the women grinding at the mill. I have seen a group of women grinding together, keeping time to a chant, or gossiping and chatting, while the younger lassies would jest and chatter and make the pine woods merry with their laughter.

Mothers carry their babes curiously in baskets. They make a wicker board by platting willows and sew a buckskin cloth to either edge, and this is pulled in the middle so as to form a sack closed at the bottom. At the top they make a wicker shade, like "my grandmother's sunbonnet," and wrap-

---

[1] Probably the *tunas,* or fruits, of the prickly pear (*Opuntia* spp.).

ping the little one in a wild-cat robe, place it in the basket, and this they carry on their backs, strapped over the forehead, and the little brown midgets are ever peering over their mothers' shoulders. In camp, they stand the basket against the trunk of a tree or hang it to a limb.

There is little game in the country; yet they get a mountain sheep now and then or a deer, with their arrows, for they are not yet supplied with guns. They get many rabbits, sometimes with arrows, sometimes with nets. They make a net of twine, made of the fibers of a native flax. Sometimes this is made a hundred yards in length, and is placed in a half circular position, with wings of sage brush. They have a circle hunt, and drive great numbers of rabbits into the snare, where they are shot with arrows. Most of their bows are made of cedar, but the best are made of the horns of mountain sheep. These are taken, soaked in water until quite soft, cut into strips, and these glued together; and such bones are quite elastic.

During the autumn, grasshoppers are very abundant. When cold weather comes, these insects are numbed and can be gathered by the bushel. At such a time they dig a hole in the sand, heat stones in a fire near by, put some in the bottom of the hole, put on a layer of grasshoppers, then a layer of hot stones, and continue this until they put bushels on to roast. There they are left until cool, when they are taken out thoroughly dried, and ground into meal. Grasshopper-gruel or grasshopper cake are articles of common food.[2]

Indians of the same race, farther to the east in the Rocky Mountains,[3] obtain grasshoppers in great quantities, collected in another manner. Late in the season, when the wings of the grasshoppers are fully fledged, they rise in vast numbers like clouds in the air and drift eastward with the upper currents. Coming near to these high snowclad mountains, they are often chilled, and fall on the great sloping sheets of snow that are spread over the mountain-sides, and tumble down these snow banks in vast numbers until

---

[2] The seemingly earlier version in *Exploration* said "Grasshopper gruel and grasshopper cake are a great treat." One wonders if the editor at *Scribner's* drew the line at such an affront to American sensibility.

[3] Probably Utes.

they are collected at the foot in huge wind-rows, often containing hundreds, thousands, tens of thousands of bushels. Here the grizzly bears come and gorge themselves on this dainty food. There the Indians come and kill the grizzly bears and gather grasshoppers. Grasshopper pudding, with bear-grease sauce, is considered a great delicacy.

Their lore consists of a mass of traditions or mythology. It is very difficult to induce them to tell it to white men; but the old Spanish priests, in the days of the conquest of New Mexico, spread among the Indians of this country many Bible stories, which the Indians are usually willing to tell.[4] It is not always easy to recognize them. When a Bible story is grafted upon a pagan legend, it becomes a curious plant, and sends forth many shoots, quaint and new. May be, much of their added quaintness is due to the way in which they were told by the "fathers." But in a confidential way, when you are alone, or when you are admitted to their camp-fire on a winter night, you will hear the stories of their mythology. I believe that the greatest mark of friendship or confidence that an Indian can give, is to tell you his religion. After one has so talked with me I should always trust him; and I felt on very good terms with these from the night on which we heard the legend of One-Two.[5]

That evening, the Shi'wits, for whom we have sent, came in, and after supper we held a long council: a blazing fire was built, and around this we sat: the Indians living here, the Shi'wits, Jacob Hamblin, and myself.[6]

This man, Hamblin, speaks their language well, and has a great influence over all the Indians in the region round about. He is a silent, reserved man,

[4]Franciscans Atanasio Domínguez and Silvestre Vélez de Escalante passed through the plateau country in 1776; after them, few Spaniards and fewer priests would have had much contact with the Southern Paiute.

[5]On that night, described earlier in the article from which this excerpt is taken, To-mo-ro-un-ti-kai, the leader of a Uinkaret band, whom Powell plied with coffee and tobacco, related a long mythological tale, even though the winter season for such tale telling was months away.

[6]Guided by the Kaibab Paiute Chu-ar, Powell and Hamblin had encountered the Uinkarets, who in turn contacted the Shivwits. Each group occupied a different territory, which was loosely defined by a plateau and its surrounding canyons.

and when he speaks it is in a slow, quiet way that inspires great awe. His talk is so low that they had to listen attentively to hear, and they sat around him in death-like silence. When he finished a measured sentence the chief repeated it and they all gave a solemn grunt. But first, I filled my pipe, lit it, and took a few whiffs, then passed it to Hamblin; he smoked, and gave it to the man next, and so it went around. When it had passed the chief, he took out his own pipe, filled and lit it, and passed it around after mine. I could smoke my own pipe in turn, but when the Indian pipe came round, I was nonplussed. It had a large stem, which, at some time, had been broken, and now there was a buckskin rag wound around it and tied with sinew, so that the end of the stem was a huge mouthful.[7] To gain time, I refilled it, then engaged in very earnest conversation, and all unawares I passed it to my neighbor unlighted.

I told the Indians that I wished to spend some months in their country during the coming year and that I should like them to treat me as a friend. I did not wish to trade, did not want their lands. Heretofore I had found it very difficult to make the natives understand my object, but the gravity of the Mormon missionary helped me much. I told them that all the great and good white men are anxious to know very many things, that they spend much time in learning, and that the greatest man is he who knows the most; that they want to know all about the mountains, and the valleys, the rivers, and the canyons, the beasts, and birds, and snakes. Then I told them of many Indian tribes, and where they live; of the European nations; of the Chinese, of Africans, and all the strange things about them that came to my mind. I told them of the ocean, of great rivers and high mountains, of strange beasts and birds. At last I told them I wished to learn about their canyons and mountains, and about themselves, to tell other men at home, and that I wanted to take pictures of everything, and show them to my friends. I told them that I could stay but a short time with them then, but that I should be back again and stay with them many months. All this occupies much time, and the matter and manner make a deep impression.

---

[7]In his description of the pipe in the *Exploration*, Powell adds that it was "exceedingly repulsive."

Then their chief replied: "Your talk is good, and we believe what you say. Your heart is good. We believe in Jacob, and look upon you as a father. When you are hungry, you may have our game. You may gather our sweet fruits. We will give you food when you come to our land. We will show you the springs and you may drink; the water is good. We will be friends, and, when you come, we shall be glad. We shall tell the Indians who live on the other side of the great river that we have seen Ka-pu-rats, and he is the Indians' friend. We shall tell them he is Jacob's friend. We are very poor. Look at our women and children; they are naked. We have no horses; we climb the rocks and our feet are sore. We live among rocks, and they yield little food and many thorns. When the cold moons come, our children are hungry. We have not much to give; you must not think us mean. You are wise; we have heard you tell strange things. We are ignorant. Last year we killed three white men. Bad men said they were our enemies. They told great lies. We thought them true. We were mad; it made us big fools. We are very sorry. Do not think of them; it is done; let us be friends. We are ignorant—like little children, in understanding, compared with you. When we do wrong, do not get mad and be like children too.

"When white men kill our people, we kill them. Then they kill more of us. It is not good. We hear that the white men are a great number. When they stop killing us, there will be no Indian left to bury the dead. We love our country; we know not other lands. We hear that other lands are better; we do not know. The pines sing and we are glad. Our children play in the warm sand; we hear them sing and are glad. The seeds ripen, and we have to eat, and we are glad. We do not want their good lands; we want our rocks and the great mountains where our fathers lived. We are very poor; we are very ignorant; but we are very honest. You have horses and many things. You are very wise; you have a good heart. We will be friends. Nothing more have I to say."

Ka-pu-rats is the name by which I am known among the Utes and Shoshones, meaning "no right arm."[8] There was much more repetition

---

[8] "Ka-pu-rats" is said by some to be the origin of the name of the Kaiparowits Plateau. However, it is more likely that the plateau received its name from a Paiute band that at this time was already known as the Kaiparowits (or a close variant).

than I have given, and much more emphasis. After this a few presents were given, we shook hands, and the council broke up.

Mr. Hamblin then fell into conversation with one of the men, and held him until the others had left, and learned more of the particulars of the death of the three men. It seems that they came upon the Indian village almost starved and exhausted with fatigue. They were supplied with food, and put on their way to the settlements. Shortly after they had left, an Indian from the east side of the Colorado arrived at their village and told them about a number of miners having killed a squaw in drunken brawl, and no doubt these were the men. No person had ever come down the cañon; that was impossible; they were trying to hide their guilt. In this way he worked them into a great rage; they followed, surrounded the men in ambush, and filled them full of arrows.

That night I slept in peace, although these murderers of my men, and their friends, the U-in-ka-rets, were sleeping not 500 yards away. While we were gone to the cañon,[9] the pack train, and supplies enough to make an Indian rich beyond his wildest dreams, were all left in their charge, and all was safe; not even a lump of sugar was pilfered by the children.

So strangely do virtues and vices grow together in the human heart; here were savages faithful to trust on one day, who, but a short time before, had been guilty of horrible, though unconsidered crime. He who sees only their crimes, and studies the history of their barbarities as it has been recorded for the past three or four centuries, can see in the Indian race only hordes of demons who stand in the way of the progress of civilization, and who must, and ought to be destroyed. He who has a more intimate knowledge of Indian character and life sometimes forgets their baser traits, and sees only their virtues, their truth, their fidelity to a trust, their simple and innocent sports, and wonders that a morally degenerate, but powerful civilization should destroy that primitive life.

Social problems are so complex that few are willing or able to comprehend all the factors, and so the people are divided into two great parties,

---

[9] Powell and a few others had descended from the rim to the Colorado River on September 17 and returned the next day. The council with the Shivwits took place on September 19.

one crying for blood, and demanding the destruction of the Indians, the other begging that he may be left in his aboriginal condition, and that the progress of civilization may be stayed. Vain is the clamor of either party; the march of humanity cannot be stayed; fields must be made, and gardens planted in the little valleys among the mountains of that Western land, as they have been in the broader valleys and plains of the East, and the mountains must yield their treasure of ore to the miner, and, whether we desire it or not, the ancient inhabitants of the country must be lost; and we may comfort ourselves with the reflection that they are not destroyed, but are gradually absorbed, and become a part of more civilized communities.

The Ancient Province of Tusayan

Powell now turns his attention eastward, beginning a journey that will carry him from Kanab across the Colorado River and onward to the ancient mesa-top villages of the Hopi.

It was the 23d of September. We had made an overland trip from Salt Lake City to the Grand Cañon of the Colorado, and were now on the bank of the Kanab, on the way back to the rendezvous camp at the upper springs of the river, which was yet about forty miles away, and which was to be our point of departure for the "Province of Tusayan."[1]

Since the exploration was made of which I am giving a general account in these papers, this stream has been carefully surveyed. Let me describe it. It is about eighty miles long, and in its course runs through three cañons which we have called the upper, middle, and lower Kanab cañons. Along its upper course for about a dozen miles it is a permanent stream, but just before entering the first cañon the water is lost in the sands. It is only in seasons of extreme rains that the water flows through this cañon, which is dry sometimes for two or three years in succession. The bed of the stream is usually dry between the upper and middle cañon. At the head of the middle

---

Powell, "The Ancient Province of Tusayan," *Scribner's Monthly* 11 (1896): 193–213. The article is reproduced here in its entirety.

---

[1] Documents of the sixteenth-century Coronado expedition refer to the Hopi villages variously as *Tucano, Tusayán, Tuçayan,* and *Tuzán.* After Coronado this group of names ceased to appear in written accounts until Powell and others revived it as *Tusayan* (Connelly, "Hopi Social Organization," 551).

cañon the water again gushes out in springs, and there is a continual stream for a dozen miles. About five miles below this cañon the water again sinks in the sands, and for ten miles or more the stream is lost, except in times of great rains, as above. This usually dry course of the stream is along a level plain where the sands drift, and sometimes obliterate all traces of the water-course. At the head of the lower cañon springs are again found, and the waters gather so as to form, in most seasons, a pretty little creek, though, in seasons of extreme drought, this is dry nearly down to the Colorado; but, in seasons of great rains, immense torrents roll down the gorge. Thus we have a curiously interrupted creek. In three parts of its course it is a permanent stream, and in two parts intermittent.

The point where we struck the Kanab was at the foot of the middle cañon where the flow of waters is perpetual, and just there we found a few pioneers of a Mormon town, to be called, after the stream, Kanab.[2] At that time these people were living in what they called a "fort"—that is, several little cabins had been built about a square, the doors and windows opening toward the plaza, the backs of their houses connected by a rude stockade made of cedar poles planted on end. This "fort" was intended for defense against the Indians.

The way in which these Mormon settlements are planted is very inter-esting. The authorities of the "Church of Jesus Christ of Latter-Day Saints" determine to push a settlement into a new region. The country is first explored and the site for a town selected, for all settlements are made by towns. The site, having been chosen, it is surveyed and divided into small lots of about an acre, with outlying lots of five or ten acres. Then a number of people are selected "to go on mission," as it is termed. The list is made out in this way: The President of the Church,[3] with his principal bishops and other officers, meet in consultation, and select from the various settle-ments throughout the territory persons whom they think it would be well to send to the new place. Many are the considerations entering into this selec-

---

[2] *Kanab* appears to have derived from a Paiute place name.
[3] Brigham Young, 1801–1877.

tion. First, it is necessary to have an efficient business man, one loyal to the Church, as bishop or ruler of the place, and he must have certain counselors; it is necessary, too, that the various trades shall be represented in the village—they want a blacksmith, shoemaker, etc. Again, in making the selection, it is sometimes thought wise to take men who are not working harmoniously with the authorities where they are residing; and thus they have a thorough discussion of the various parties, and the reasons why they are needed here and there; but at last the list is made out. The President of the Church then presents these names to the General Conference of the Church for its approval, and that body having confirmed the nominations (and perhaps there is no instance known where a nomination is not confirmed), the people thus selected are notified that at a certain time they are expected "to go on a mission" to establish a new town. Sometimes a person selected, feeling aggrieved with the decision of the Church, presents his reasons to the President for wishing to remain, and occasionally such a person is excused, but the reasons must be very urgent. So far as my observation goes, there is rarely any determined opposition to the decision of the Conference.

So the people move to their new home. Usually there are four lots in a square, and four persons unite to fence the same, each receiving a garden. The out-lots are fenced as one great farm. The men, living in covered wagons or tents, or having built cabins or other shelter for themselves, set to work under the bishop or one of his subordinates to fence the farm, and make the canals and minor water-ways necessary to the irrigation of the land. The water-ditch and fence of the farm are common property.[4] As soon as possible a little store is established, all of the principal men of the community taking stock in it, usually aided more or less by "Zion's Cooperative Mercantile Institution," the great wholesale establishment in Salt Lake City. In the same way saw-mills and gristmills are built.

---

[4]Powell's observation of Mormon and, to a lesser degree, Hispanic communalism in building and maintaining irrigation infrastructure had a profound influence on his later recommendations for the settlement of the West.

Such is a brief outline of the establishment of a Mormon town; in like manner, all of the towns throughout the territory of Utah have branched out from the original trunk at Salt Lake City, so that they are woven together by a net-work of communal interest.

The missionary, Jacob Hamblin, who was traveling with us, came here two or three years ago and established himself in a little cabin, about which during the greater part of each season a few Indians were gathered. When we came to the place, we found the men at work cutting and hauling hay, while a number of squalid Indians were lounging in the "fort," and many children of white and Indian breed were playing in the meadow. Such a community is a strange medley of humanity. There are no physicians here, but the laying on of hands by the elders is frequently practiced, and every old man and woman of the community has some wonderful cure—a relic of ancient sorcery. Almost every town has its astrologer, and every family one or more members who see visions and dream dreams. Aged and venerable men, with solemn ceremony, are endowed by the Church with the power of prophecy and the gift of blessing. So the grandfather recounts the miracles which have been performed by the prophets; the grandmother tells of the little beast that has its nest in the heart, and when it wanders around toward the lungs causes consumption; the mother dreams dreams; the daughter consults the astrologer and the son seeks for a sign in the heavens. At every gathering for preaching on a Sunday morning, or dancing on a weekday night, a prayer is offered. When they gather at table, thanks are rendered to the Giver of Bounties, and on all occasions, and in the most earnest manner, when a stranger is met, the subject of miracles, the persecution of the saints, and the virtue and wisdom of polygamy are discussed.

Good roads are built to every settlement, at great expense and with much labor. The best agricultural implements are found on the farms, and the telegraph clicks in every village. Altogether, a Mormon town is a strange mixture of Oriental philosophy and morals, primitive superstitions and modern inventions.

I must not fail to mention here the kind treatment which I have almost invariably received from the people living in the frontier settlements of Southern Utah.

At Kanab, the party divided, Mr. Hamblin, with one man, going to Tokerville—a settlement about fifty miles to the north west—for the purpose of procuring some additional supplies. With the remainder of the party I proceeded up the Kanab. The trail was very difficult; it was impossible to climb the cliffs and go over the plateau with our animals, and we had to make our way up the cañon. In many places the stream runs over beds of quicksand, sweeping back and forth in short curves from wall to wall, so that we were compelled to ford it now and then; again, there is a dense undergrowth, and, at many places, the stream is choked with huge boulders which have fallen from the cliffs. The plateau, or terrace, through which this cañon is cut, slopes backward to the north, and, by ascending the stream, we at last reached its summit, and found it covered with a sea of drifting sands, golden and vermilion; so we named it Sand-Dune Plateau. Just before us, there was another line of cliffs—a great wall of shining white sandstone, a thousand feet high.

We soon entered another cañon, but this was dry. At some very late geological period a stream of lava has rolled down it, so that we had to pass over beds of black clanking basalt.[5] At night, having emerged from the upper cañon, we found the Kanab a living stream once more, and camped upon its bank.

The next day we passed up the beautiful valley for ten miles, and arrived at the rendezvous camp. Here I was to wait for a few days for Mr. Hamblin's arrival. I kept the Indians and one white man with me, and Mr. Nebeker, with the remainder of the party and a single Indian guide, started for the Colorado River, at the mouth of the Paria, by a well traveled Indian trail. We had brought a quantity of lumber to this point with wagons, for the purpose of building a ferry-boat on the Colorado. These boards were cut into short pieces and packed on mules, and Mr. Nebeker was to push on to the river, construct the boat, get the train across, and have everything in readiness, on

---

[5] "Clanking basalt" is not a geological term. Powell is simply referring to the sound that loose basaltic rocks make when hard objects such as horseshoes or other rocks strike them.

the opposite side of the river, by the time of our arrival. My purpose was to demonstrate the practicability of this route to the river, then to cross at the mouth of the Paria, and proceed thence to the "Province of Tusayan," in north-eastern Arizona.

The Indians we had with us were not acquainted with the country beyond the river, and it was necessary to obtain some new aids, so I sent Chu-ar to the Kaibab Plateau, a hundred miles to the south-east, with instructions to collect the Indians who inhabit that region at a designated spring, and hold them until my arrival.[6]

I waited a week in the upper valley of the Kanab, the time being chiefly spent in talking with the Indians, and trying to learn something of their language. By day the men hunted, and the women gathered berries and the other rich fruits that grow in that country, and at night they danced. A little after dark a fire was kindled, and the musicians took their places. They had two kinds of instruments. One was a large basket tray, covered with pitch inside and out, so as to be quite hard and resonant; this was placed over a pit in the ground, and they beat on it with sticks. The other was a primitive fiddle, made of a cedar stick, as large around as my wrist and about three feet long; this was cut with notches about three inches apart. They placed one end on a tray arranged like the one just described, placed the other end against the stomach, and played upon the fiddle with a pine-stick bow, which was dragged up and down across the notches, making a rattling, shrieking sound. So they beat their loud drum and sawed their hoarse fid-

---

[6] Chu-ar, short for Chuar'-ruumpeak, was also a Kaibab but from the Kanab area. Evidently he was unfamiliar with lands south and east of the Colorado (see *Exploration*, 290). Elsewhere, Powell expresses profound admiration for Paiute geographic knowledge:

There is not a trail but what they know; every gulch and every rock seems familiar. I have prided myself on being able to grasp and retain in my mind the topography of a country; but these Indians put me to shame. My knowledge is only general, embracing the more important features of a region that remains as a map engraved on my mind; but theirs is particular. They know every rock and every ledge, every gulch and canyon, and just where to wind among these to find a pass; and their knowledge is unerring. They cannot describe a country to you, but they can tell you all the particulars of a route. (*Exploration*, 299–300)

dle for a time until the young men and maidens gathered about and joined in a song:

Ki-ap-pa tu-gu wun,
Pi-vi-an-na kai—va.
(Friends, let the play commence;
All sing together.)

Gradually they formed a circle, and the dance commenced. Around they went, old men and women, young men and maidens, little boys and girls, in one great circle, around and around, all singing, all keeping time with their feet, pat, pat, pat, in the dust and sand; low, hoarse voices; high, broken, screaming voices; mellow, tender voices; but louder than all, the thump and screech of the orchestra.

One set done, another was formed; this time the women dancing in the inner circle, the men without. Then they formed in rows, and danced, back and forth in lines, the men in one direction, the women in another. Then they formed again, the men standing expectant without, the women dancing demurely within, quite independent of one another, until one maiden beckoned to a lover, and he, with a loud, shrill whoop, joined her in the sport. The ice broken, each woman called for her partner; and so they danced by twos and twos, in and out, here and there, with steadily increasing time, until one after another broke down and but three couples were left. These danced on, on, on, until they seemed to be wild with uncontrollable motion. At last one of the couples failed, and the remaining two pattered away, while the whole tribe stood by shouting, yelling, laughing, and screaming, until another couple broke down, and the champions only remained. Then all the people rushed forward, and the winning couple were carried and pushed by the crowd to the fire. The old chief came up, and on the young man's head placed a crown of eagle feathers. A circlet of braided porcupine quills was placed about the head of the maiden, and into this circlet were inserted plumes made of the crest of the quail and the bright feathers of the humming-bird.

On the first of October, Mr. Hamblin having returned from Tokerville, we started for the Kaibab Plateau to meet the Indians, as had been arranged with Chu-ar. That night we camped in the cañon of the Skoom-pa.[7] This is really a broad cañon valley, the walls of which are of red sandstone. On the lower reaches of these walls, near some springs, there are many hieroglyphics, some of them so high up as to be beyond reach, in the present condition of the talus at the foot of the cliffs.

The next day our course was through barren sage plains until, about four o'clock, we came to the foot of the Kaibab Plateau, and went up a gulch, where we hoped to find water in a limestone pocket, but were disappointed. This compelled us to continue our journey long into the night. The direction traveled was now to the south, and our way was up a long cañon valley, with high mountains on either side. At last we reached a spring, and camped.

Three hours' travel the next morning brought us to the spring at which we were to meet the Indians, but none were seen. High up on the mountain to the east was a signal smoke, which we understood, by previous arrangement, meant that we were to cross the Kaibab Plateau. We stayed in camp the remainder of that day to rest.

The next day we started early, climbing to the summit of the plateau, more than two thousand feet up a long, rocky gulch; then through a forest of giant pines, with glades here and there, and now and then a lake. Occasionally a herd of deer was started. In this upper region, eight thousand feet above the level of the sea, even the clouds of northern Arizona yield moisture sufficient for forest growth and rich meadows. At dusk we descended from the plateau on the eastern side, found a spring at its foot, and camped.

The next day we crossed a broad valley to the foot of the line of Vermilion Cliffs, and at two o'clock reached the designated spring, where we found our Indians. They had already arranged that Na-pu and To-ko-puts

---

[7] Skutumpah Creek lies ten miles northeast of the town of Kanab. The modern name derives from the terms Powell used: *skoom*, "rabbitbrush" + *pa*, "creek or water source" = "the creek where rabbitbrush grows."

(Old Man and Wild Cat) should be our guides from the Colorado River to the "Province of Tusayan."

During the evening I was very much interested in obtaining from them a census of their little tribe. They divided the arithmetic into parts, each of four men taking a certain number of families. Each sat down and counted on his fingers and toes the persons belonging to the families allotted to him, going over them again and again until each finger and toe stood in his mind for an individual. Then he would discuss the matter with other Indians, to see that all were enumerated, something like this: "Did you count Jack?" "Yes; that finger stands for Jack." "Did you count Nancy?" "Yes; that toe is Nancy." Each of the census takers becoming satisfied that he had correctly enumerated his portion, he procured the number of sticks necessary to represent them, and gave them to me. Adding the four together, I had the census of the tribe—seventy-three. Then I set them to dividing them severally into groups of men, women and children, but this I found a hard task. They could never agree among themselves whether certain persons should be called children, or not; but, at last, I succeeded in obtaining the number of males and females.

The next morning I distributed some presents of knives, tobacco, beads, and other trinkets, and we pushed on toward the Colorado River. We found a difficult trail, having to cross the heads of many abrupt, but not very deep cañons. Down and up we climbed all day long, winding about here and there, and always among the rocks, until at night we joined our party at the mouth of the Paria, and were ferried over to their camp.[8]

---

[8] Powell glosses over the achievement of Nebeker and his party (who preceded Powell and Hamblin to the river from Kanab) in building the needed ferryboat and placing it in operation. It is not clear whether John Doyle Lee was in this first party to use "Lee's Ferry," but use of the boat soon passed to him. Lee had participated in the Mountain Meadow Massacre of 1857, in which a Mormon militia slaughtered a non-Mormon emigrant wagon train. He may have hoped that the solitude of the Pariah would cause the world to forget about him, but it did not. In 1874 he was arrested for his role in the massacre, and in 1877 he was executed, the only Mormon to be punished for the incident. Emma Lee, the seventeenth of his nineteen wives, operated the ferry until 1879, and the ferry remained the only reliable crossing on the river for a vast distance until 1928. Lees Ferry (written without an apostrophe) at the confluence of the Paria and Colorado lies a mile above "Lee Ferry," which marks the official point of division between the upper and lower basins of the Colorado River.

Early the next morning I climbed the Vermilion Cliffs. This great escarpment or wall of flaring red rock in a general direction faces south, from Saint George on the Rio Virgen to a point many miles east of the Colorado River, a distance of more than three hundred miles as we follow the meandering line. There is a deep re-entrant angle at the mouth of the Paria, where I climbed. Standing on an elevated point on the cliffs, and looking southward, I could see over a stretch of country that steadily rose in the distance until it reached an altitude far above even the elevated point of observation; and then, meandering through it to the south, the gorge in which the river runs, everywhere breaking down with a sharp brink, and the summits of the walls appearing to approach until they merged in a black line; and could hardly resist the thought that the river burrowed into, and was lost in, the great inclined plateau. This gorge was Marble Cañon, described in a previous article.[9]

While I was climbing, the train pushed on, in a direction a little to the east of south, along the foot of the Vermilion Cliffs.[10] By mid-afternoon I overtook it. The trail by which we were traveling led up into a deep gulch, and we came to a clear, beautiful spring, gushing from beneath a rock a thousand feet high. Here was indeed "the shadow of a great rock in a weary land," and here we camped for the night. All about us were evidences of an ancient town or hamlet, foundation walls of houses half buried in debris, fragments of pottery painted with rude devices, and picture writings etched on the Cliffs.

----

[9] Marble Canyon stretches south along the Colorado River roughly from Lees Ferry to the mouth of the Little Colorado River, after which the main stem of the Colorado soon bends west. As mentioned earlier, Powell named the canyon for its immense vertical walls of limestone, which are stained red from minerals in an overlying formation. Test drilling for a high dam midway through the canyon began in the 1960s, but environmental opposition to the drowning of Marble Canyon ultimately defeated the project.

[10] The cliffs on the east side of the river are today called Echo Cliffs. A short distance upstream of the point where the river cuts through the joint structure of the Vermilion and Echo cliffs is the site of Glen Canyon Dam, which was completed in 1964. The reservoir impounded by the dam is named for Powell (detractors call it Lake Foul); it stretches upstream for 186 miles, inundating Glen Canyon and a portion of the lower San Juan River.

For another day, our journey was at the foot of the Vermilion Cliffs, in a direction a little east of south, over naked hills of sand and marls, where we found briny springs occasionally, but no fresh water, and no grass; a desert, but a painted desert; not a desert plain, but a desert of hills, and cliffs, and rocks—a region of alcove lands. At night we found a little water, in a basin or pocket, a mile from the trail.[11]

The next day we went to the top of the mesa by climbing the cliffs, and found a billowy sea of sand-dunes. The line of cliffs, separating the mesa above from the deeply gulch-carved plain below, is a long irregular and ragged region, higher by many hundred feet than the general surface of the mesa itself. On the slope of this ridge, facing the mesa, there is a massive homogeneous sandstone, and the waters, gathering on the brink of the ridge and rolling down this slope, have carved innumerable channels; and, as they tumble down precipitously in many places, they dig out deep pot-holes, many of them holding a hundred or a thousand barrels of water. Among these holes we camped, finding a little bunch grass among the sand-dunes for our animals. We called this spot the Thousand Wells.

Leaving the wells, we trudged for a day among the sand-dunes, and at night found a deep cave in a ledge of rocks, and, in the farther end of the cave, a beautiful lake. Here our Indian guides discovered evidences that led them to believe that our track was followed by some prowling Indians. In the sands about the cave were human tracks; these our guides studied for some time, and, while they were thus engaged, the white men of the party also talked the matter over, with a little anxiety, for we were now in the country of the Navajos, who had lately been making raids on the Mormon settlements, stealing horses and cattle, and occasionally killing a man, and we feared that they might be following us. In talking with Na-pu, he assured me that they were not Navajos, but doubtless belonged to a band of Indians known to our tribe as Kwai-an-ti-kwok-ets,[12] or "Beyond the river people,"

---

[11] Natural stone pools that collect rainwater or snowmelt—called water pockets or, in Spanish, *tinajas*—are an important source of water in desert country.
[12] Another Southern Paiute band.

and were their friends. His reasons were these: The tracks which they made in the sand were evidently made with moccasins having projecting soles, like those worn by our Indians and their friends, while the moccasins worn by the Navajos have no such projecting soles. Again, one of the tracks, as he showed me, was made by a lame man, with his right leg shortened, so that he could only walk on the toes of that foot, and this, he said, was the case with the chief of the Kwai-an-ti-kwok-ets. Again, said Na-pu, they would not have walked in places where their tracks would be exposed had they been unfriendly. The conclusion he came to was that they were anxious to see us, but were afraid we had hostile intentions. I directed him to go to an eminence near by and kindle a signal-fire. This he did, and, an hour afterward, three Indians came up. We sat, and talked with them until midnight; but they seemed surly fellows, and the conversation was not satisfactory to me. At last they left us; but, for fear they would attempt to steal some of our animals, I had the latter collected, and, finding that we should lose our rest by watching them, I concluded that we might as well continue our journey. So, at two o'clock, everything was packed, we took breakfast, and started, finding our way across the country in the direction we wished to travel, guided by the stars.

Na-pu, the old Indian guide, usually rode with me, while To-ko-puts remained with the men who were managing the pack train. The old man was always solemn and quite reticent, but that day I noticed that he was particularly surly. At last I asked him why. "Why you never call me 'a brick'?" he replied. The answer, of course, astonished me; but, on thinking, and talking with him a little further, I understood the matter. For the previous two or three days we had been quite anxious about water, and the other man, To-ko-puts, when camping time came, usually ran ahead after consulting with Na-pu; finding the watering-place, he would kindle a signal-smoke for us to come on. On arriving, the men, pleased with the Indian's success, would call him "a brick," and thus, it seemed to the old man, that the younger took all the honors away from him; and he explained to me that in his boyhood he had lived in this country, and that it was his knowledge that guided To-ko-puts altogether. I soothed his wounded feelings in this way. He could see that To-ko-puts laughed and talked with the "boys," and was a boy with the rest, but that he (Na-pu) and I were old men, and I recognized his wisdom in the matter. This satisfied him, and ever after that he seemed to be at great pains to talk no more with the younger members of the party, but always came to me.

At ten o'clock we came in sight of a deep depression made by the Mo-an-ka-pi, a little stream which enters the Colorado Chiquito.[13] Before us, two or three miles, was the meandering creek, with a little fringe of green willows, box-elders, and cotton-woods; from these, sage plains stretched back to the cliffs that form the walls of the valley. These cliffs are rocks of bright colors, golden, vermilion, purple and azure hues, and so storm-carved as to imitate Gothic and Grecian architecture on a vast scale. Outlying buttes were castles, with minaret and spire; the cliffs, on either side, were cities looking down into the valley, with castles standing between; the inhabitants of these cities and castles are a million lizards: great red and black lizards, the kings of nobles; little gray lizards, the common people, and here and there a priestly rattlesnake.

We went into camp early in the day, and, with Mr. Hamblin, I started away to the north to visit what had often been described to me as an artificial wall extending across the country for many miles, and one, two, or three hundred feet high; it was claimed, further, that the blocks of which the wall was composed had been carried from a great distance, from the fact that they were not rocks found in that region, but only to the northwest, among the mountains. We were well mounted and rode across the country at a good gallop, for nearly a score of miles, when we came to the wonderful wall, the fame of which had spread among all the Mormon towns to the west. We found it in fact to be an igneous dike, the blocks composed of columnar basalt.[14] In the joints between the blocks there is

---

[13] Moenkopi Wash and the Little Colorado River, respectively.
[14] Such a dike may form in the following manner: Basaltic lava flows down a narrow valley and cools, taking on the shape of the valley bottom. Because of the physical properties of its crystalline structure, the basalt fractures vertically, giving it a columnar appearance. Through time, the surrounding landscape erodes away, exposing the more resistant basalt as a wall of rock.

often an accumulation of a whitish mineral, having the effect, in a rude way, of suggesting mortar. It is not, in fact, a single dike, but a number, radiating from a common center, a great mass of basalt, forming quite a large hill, which the Indians call Kwi-pan-chom, a word signifying "axe hill," for here the Indians of the adjacent country obtain the material for their axes.

Late in the evening a number of Navajo Indians rode up to our camp. One of them could speak a little Spanish or Mexican patois. After a little conversation, they concluded to stay with us during the night, tempted, perhaps, by the sight and odor of biscuits and coffee. They were fine-looking fellows, tall and lithe, with keen eyes, sharp features, and faces full of animation. After supper, our new friends and the Kai-bab-it guides sat down for a conference. It was very interesting to observe their means of communicating thought to each other. Neither understood the oral language of the other, but they made maps with their fingers in the sand describing the whereabouts of the several tribes, and seemed to have a great deal of general discussion by means of a sign language. Whenever an Indian's tongue is tied he can talk all over; and so they made gestures, struck attitudes, grunted, frowned, laughed, and altogether had a lively time.

The next morning a Navajo boy offered to go with us to Oraibi, for the purpose of showing us the shortest way. After dinner, we descended from the table-land on which we had been riding, into a deep valley, and, having crossed this, commenced to ascend a steep rocky mesa slope by a well-worn trail, and were surprised, on approaching the summit, to find the slope terraced by rude masonry, which had evidently been made with great labor. These terraces, two or three acres in all, were laid out in nice little gardens, carefully irrigated by training water from a great spring in little channels among the garden plats [sic]. Here we found a number of men, women and children from the town of Oraibi gathering their vegetables. They received us with hearty welcome and feasted us on melons. Then we pushed on in company with our new-found friends, rather a mixed crowd now—white men, Kai-bab-its, Navajos, and Shi-nu-mos.[15]

A little before sundown we arrived at Oraibi, the principal town in the "Province of Tusayan," and were met by some of the men, who, at our request, informed us where we could find a good camp. Later in the evening, the chief, who was absent when we arrived, came to camp, and placed our animals in the charge of two young men, who took them to a distance from the town and herded them for the night.

The "Province of Tusayan" is composed of seven towns—Oraibi, Shi-pau-a-luv-i, Mishong-ini-vi, Shong-a-pa-vi, Te-wa, Wol-pi, and Sichoam-a-vi.[16] The last three are known as the Moqui Towns.[17]

We remained nearly two months in the province, studying the language and customs of the people;[18] and I shall drop the narrative of travel, to describe the towns, the people, and their daily life.

Oraibi and the three Moqui towns are greatly dilapidated, and their original plans are not easily discovered. The other three towns are much better preserved. There are now about two thousand seven hundred inhabitants

---

[15] Powell's reference to Shinumos is not entirely clear. Shinumo Wash enters the Colorado from the east and drains the country in which he encountered the three "surly" Paiutes who he feared might steal the horses. Perhaps these three or others from that area had tagged along with his group.

[16] Powell gives *Shi-pau-a-luv-i* in several spellings. Today it is *Sipaulovi. Mishong-ini-vi* usually is given as *Mishongnovi, Shong-a-pa-vi* as *Shungopavi, Wol-pi* as *Walpi,* and *Sichoam-a-vi* as *Sichomovi;* Tewa, where Tewa refugees from the Rio Grande resettled after the Pueblo Revolt of 1680, is better known as Hano.

[17] *Moqui* is derived for a Hopi term of self-description. Mispronounced, it becomes indistinguishable from the Hopi term for "dies" or "is dead," which was exceedingly offensive to the Hopi. Anthropologist Jesse Fewkes led an effort to substitute the word *Hopi,* and in 1923 the term *Moqui* was dropped from government use (Connelly, "Hopi Social Organization," 551).

[18] In *Exploration,* where this material is presented in somewhat different form, Powell writes, "And so the days pass and the weeks go by, and we study the language of the people and record many hundreds of their words and observe their habits and customs and gain some knowledge of their mythology, but above all do we become interested in their religious ceremonies" (p. 338). In this way, Powell established a pattern that innumerable later anthropologists followed, often to the chagrin of the Hopi.

in the seven towns, probably but a small proportion of what they at one time contained. The towns are all built on high cliffs or rocks, doubtless for greater security against the common nomadic enemies, the Navajos on the north and Apaches on the south. Each town has a form peculiar to itself and adapted to its site—Shi-pau-a-luv-i the most regular, Oraibi the most irregular. Shi-pau-a-luv-i is built about an open-court; the exterior wall is unbroken, so that you enter the town by a covered way. Standing within, the houses are seen to be two, three, and four stories high, built in terraces—that is, the second story is set back upon the first, the third back upon the second, the fourth upon the third; the fourth or upper story being therefore very narrow. Usually, to enter a room on the first story from the court, it is necessary to climb by a ladder to the top of the story, and descend by another through a hatchway. To go up to the third or fourth story you climb by a stairway made in the projecting wall of the partition. The lower rooms are chiefly used for purposes of storage. The main assembly-room is in the second story, sometimes in the third. The rooms below are quite small, eight or ten feet square, and about six feet high. The largest room occupied by a family is often twenty to twenty-four feet long by twelve or fifteen feet wide, and about eight feet between floor and ceiling. Usually all the rooms are carefully plastered, and sometimes painted with rude devices. For doors and windows there are openings only, except that sometimes small windows are glazed with thin sheets of selenite, leaf-like crystals of gypsum.

In a corner of each principal room a little fire-place is seen, large enough to hold about a peck of wood; a stone chimney is built in the corner, and often capped outside with a pottery pipe. The exterior of the house is very irregular and unsightly, and the streets and courts are filthy; but within, great cleanliness is observed. The people are very hospitable and quite ceremonious. Enter a house and you are invited to take a seat on a mat placed for you upon the floor, and some refreshment is offered—perhaps a melon, with a little bread, perhaps peaches or apricots. After you have eaten, every thing is carefully cleaned away, and, with a little broom made of feathers, the matron or her daughter removes any crumbs or seeds which may have been dropped. They are very economical people; the desolate circumstances under which they live, the distance to the forest and the scarcity of game, together with their fear of the neighboring Navajos and Apaches, which prevents them from making excursions to a distance—all combine to teach them the most rigid economy. Their wood is packed from a distant forest on the backs of mules, and when a fire is kindled but a few small fragments are used, and when no longer needed the brands are extinguished, and the remaining pieces preserved for future use.

Their corn is raised in fields near by, out in the drifting sands, by digging pits eighteen inches to two feet deep, in which the seeds are planted early in the spring, while the ground is yet moist. When it has ripened, it is gathered, brought in from the fields in baskets, carried, by the women and stored away in their rooms, being carefully corded.[19] They take great pains to raise corn of different colors, and have the corn of each color stored in a separate room. This is ground by hand to a fine flour in stone mills, then made into a paste like a rather thick gruel. In every house there is a little oven made of a flat stone eighteen or twenty inches square, raised four or five inches from the floor, and beneath this a little fire is built. When the oven is hot and the dough mixed in a little vessel of pottery, the good woman plunges her hand in the mixture and rapidly smears the broad surface of the furnace rock with a thin coating of the paste. In a few moments the film of batter is baked; when taken up it looks like a sheet of paper.[20] This she folds and places on a tray. Having made seven sheets of this paper bread from the batter of one color and placed them on the tray, she takes batter of another color, and, in this way, makes seven sheets of each of the several colors of corn batter.

They have many curious ways of preparing their food, but perhaps the daintiest dish is "virgin hash." This is made by chewing morsels of meat and bread, rolling them in the mouth into little lumps about the size of a horse-chestnut, and then tying them up in bits of corn husk. When a num-

---

[19] The cobs are stacked, like cordwood.
[20] Powell has described the making of a tortilla on a stone griddle. Elsewhere he writes, "The bread is a great novelty to me" (*Exploration*, 336).

ber of these are made, they are thrown into a pot and boiled like dumplings. The most curious thing of all is, that only certain persons are allowed to prepare these dumplings; the tongue and palate kneading must be done by a virgin. An old feud is sometimes avenged by pretending hospitality, and giving to the enemy dumplings made by a lewd woman.

In this warm and dry climate the people live principally out of doors or on the tops of their houses, and it is a merry sight to see a score or two of little naked children climbing up and down the stairways and ladders, and running about the tops of the houses engaged in some active sport.

In every house vessels of stone and pottery are found in great abundance. These Indian women have great skill in ceramic art, decorating their vessels with picture-writings in various colors, but chiefly black.

In the early history of this country, before the advent of the Spaniard, these people raised cotton, and from it made their clothing; but between the years 1540 and 1600 they were supplied with sheep, and now the greater part of their clothing is made of wool, though all their priestly habiliments, their wedding and burying garments, are still made of cotton.[21]

Men wear moccasins, leggings, shirts and blankets; the women, moccasins with long tops, short petticoats dyed black, sometimes with a red border below, and a small blanket or shawl thrown over the body so as to pass over the right shoulder under the left arm. A long girdle of many bright colors is wound around the waist. The outer garment is also black. The women have beautiful, black glossy hair, which is allowed to grow very long, and which they take great pains in dressing. Early in the morning, immediately after breakfast, if the weather is pleasant, the women all repair to the tops of the houses, taking with them little vases of water, and wash, comb, and braid one another's hair. It is washed in a decoction of the soap plant, a species of yucca, and then allowed to dry in the open air. The married ladies have their hair braided and rolled in a knot at the back of the head, but the maidens have it parted along the middle line above, and each lock carefully braided, or

---

[21] Sheep may actually have come to the Hopi somewhat later, after 1629, when the first Hopi mission was established at Awatovi (Brew, "Hopi Prehistory and History to 1850," 519–20).

twisted and rolled into a coil supported by little wooden pins so as to cover each ear, giving them a very fantastic appearance.

I have already said that the people are hospitable; they are also very polite. If you meet them out in their fields, they salute you with a greeting which seems to mean, "May the birds sing happy songs in your fields." They have many other greetings for special occasions. Do one a favor and he thanks you; if a man, he says, "Kwa kwa"; if a woman," Es-ka-li." And this leads me to say that there is a very interesting feature in their language found among people of the same grade of civilization in other parts of the world: many words are used exclusively by men, others by women. "Father," as spoken by a girl, is one word; spoken by a boy it is another; and nothing is considered more vulgar among these people than for a man to use a woman's word, or a woman a man's.[22]

At the dawn of day the governor of the town goes up to the top of his house and calls on the people to come forth. In a few moments the upper story of the town is covered with men, women, and children. For a few minutes he harangues them on the duties of the day. Then, as the sun is about to rise, they all sit down, draw their blankets over their heads and peer out through a little opening and watch for the sun. As the upper limb appears above the horizon every person murmurs a prayer, and continues until the whole disk is seen, when the prayer ends and the people turn to their various avocations. The young men gather in the court about the deep fountain stripped naked, except that each one has a belt to which are attached bones, hoofs, horns, or bells, which they have been able to procure from white men. These they lay aside for a moment, plunge into the water, step out, tie on their belts, and dart away on their morning races over the rocks, running as if for dear life.[23] Then the old men collect the little boys, sometimes with little whips, and compel them to go through the same exercises. When the athletes return, each family gathers in the large room for breakfast. This

---

[22] Gender-specific vocabularies are a common feature of North American Indian languages.
[23] Early morning running was required of boys in many tribes. See Peter Nabokov, *Indian Running.*

over, the women ascend to the tops of their houses to dress, and the men depart to the fields or woods, or gather in the kiva to chat or weave.

This kiva, as it is called in their own tongue, is called "Estufa" by the Spaniards, and is spoken of by writers in English as the "Sweat House." It is, in fact, an underground compartment, chiefly intended for religious ceremonies, but also used as a place of social resort. A deep pit is excavated in the shaly rock and covered with long logs, over which are placed long reeds, these, in turn, covered with earth, heaped in a mound above. A hole, or hatchway, is left, and the entrance to the kiva is by a ladder down the hatchway. The walls are plastered, little niches, or quadrangular recesses, being left, in which are kept the paraphernalia of their religious ceremonies. At the foot of the wall, there is a step, or bench, which is used as a seat. When the people assemble in the kiva, a little fire is built immediately under the hatchway, which forms a place of escape for the smoke. Here the elders assemble for council, and here their chief religious ceremonies are performed, for the people are remarkable for their piety. Some of these ceremonies are very elaborate and long. I witnessed one which required twenty-four hours for its performance. The people seem to worship a great number of gods, many of whom are personified objects, powers and phenomena of nature. They worship a god of the north, and a god of the south; a god of the east, and a god of the west; a god of thunder, and a god of rain, the sun, the moon, and the stars; and, in addition, each town has its patron deity. There seems, also, to be engrafted on their religion a branch of ancestral worship. Their notion of the form and constitution of the world is architectural; that it is composed of many stories. We live in the second. Ma-chi-ta, literally the leader, probably an ancestral god, is said to have brought them up from the lower story to the next higher, in which we now live. The heaven above is the ceiling of this story, the floor of the next. Their account of their rescue from the lower world by Ma-chi-ta is briefly as follows: The people below were a medley mass of good and bad, and Ma-chi-ta determined to rescue the former, and leave the latter behind. So he called to his friends to bring him a young tree, and, looking overhead at the sky of that lower world, the floor of this, he discovered a crack, and placed the young and growing tree immediately under it. Then he raised his hands and prayed, as did all his followers; and, as he prayed, the tree grew, until its branches were thrust through the crevice in the lower-world sky. Then the people climbed up, in one long stream; still up they came until all the good were there. Ma-chi-ta, standing on the brink of the crevice, looked down, and saw the tree filled with the bad, who were following; then he caught the growing ladder by the upper boughs, twisted it from its foundation in the soil beneath, and threw it over, and the wicked fell down in a pile of mangled, groaning, cursing humanity. When the people had spread out through this world, they found the ceiling, or sky, so low that they could not walk without stooping, and they murmured. Then Ma-chi-ta, standing in the very center of this story, placed his shoulder against the sky, and lifted it to where it now is.

Still it was cold and dark, and the people murmured and cursed Ma-chi-ta, and he said: "Why do you complain? Bring me seven baskets of cotton"; and they brought him seven baskets of cotton. And he said: "Bring me seven virgins"; and they brought him seven virgins. And he taught the virgins to weave a wonderful fabric, which he held aloft, and the breeze carried it away to the sky; and behold! it was transformed into a full-orbed moon. The same breeze also carried the flocculent fragments of cotton to the sky, and lo! these took the shape of bright stars. And still it was cold; and again the people murmured, and Ma-chi-ta chided them once more, and said, "Bring me seven buffalo robes"; and they brought him seven buffalo robes. "Send me seven strong, pure young men"; and they sent him seven young men, whom he taught to weave a wonderful fabric of the buffalo fur. And when it was done, he held it aloft, and a whirlwind carried it away to the sky, where it was transformed into the sun.

I have given but a very bare account of these two chapters in their unwritten bible—the bringing up of the people from the lower world to this, and the creation of the heavenly bodies. As told by them, there are many wonderful incidents; the travels, the wandering, the wars, the confusion of tongues, the dispersion of the people into tribes—all these are given with much circumstance.

Mu-ing-wa is the god of rain, and the ceremony of which I have made mention as lasting twenty-four hours was in honor of this god, immediately after the gathering of the harvest. A priest from Oraibi, one from Shi-pau-i-luv-i, one from Shong-a-pa-vi, together with the one from Mi-shong-i-ni-vi, gathered in the kiva at this latter place. An old woman, a grandmother, her daughter, a mother and her granddaughter, a virgin, three women in the same ancestral line, were also taken into the kiva, where I was permitted to join them. Before this I had known of many ceremonies being performed, but they had always refused me admittance, and it was only the day before, at a general council held at Oraibi, that it was decided to admit me. The men were entirely naked, except that during certain parts of the ceremony they wrapped themselves in blankets, and a blanket was furnished me at such times for the same purpose. The three women were naked, except that each had a cincture made of pure white cotton wound about the loins and decorated with tassels. Event followed event, ceremony ceremony so rapidly during the twenty-four hours, that I was not able on coming out to write a very definite account of the sacred rites, but I managed to carry away with me some things which I was afterward able to record in my notes from time to time.

I have said that the ceremony was in honor of Mu-ing-wa, the god of rain. It was a general thanksgiving for an abundant harvest, and a prayer for rain during the coming season. Against one end of the kiva was placed a series of picture writings on wooden tablets. Carved wooden birds on little wooden pedestals, and many pitchers and vases, were placed about the room. In the niches was kept the collection of sacred jewels—little crystals of quartz, crystals of calcite, garnets, beautiful pieces of jasper, and other bright or fantastically shaped stones, which, it was claimed, they had kept for many generations. Corn meal, flour, white and black sand were used in the ceremony at different times. There were many sprinklings of water, which had been previously consecrated by ceremony and prayer. Often the sand or meal was scattered about. Occasionally during the twenty-four hours a chorus of women singers was brought into the kiva, and the general ceremony was varied by dancing and singing. The dancing was performed by single persons or by couples, or by a whole bevy of women; but the singing was always in chorus, except a kind of chant from time to time by the elder of the priests. My knowledge of the language was slight, and I was able to comprehend but little of what was said; but I think I obtained, by questioning and close observation, and gathering a few words here and there, some general idea of what they were doing. About every two hours there was a pause in the ceremony, when refreshments were brought in, and twenty minutes or half an hour was given to general conversation, and I always took advantage of such a time to have the immediately preceding ceremony explained to me as far as possible. During one of these resting times I took pains to make a little diagram of the position which had been assumed by the different parties engaged, and to note down, as far as possible, the various performances, which I will endeavor to explain.

A little to one side of the fire (which was in the middle of the chamber) and near the sacred paintings, the four priests took their positions in the angles of a somewhat regular quadrilateral. Then the virgin placed a large vase in the middle of the space; then she brought a pitcher of water, and, with a prayer, the old man poured a quantity into the vase. The same was done in turn by the other priests. Then the maiden brought on a little tray or salver, a box or pottery case containing the sacred jewels, and, after a prayer, the old man placed some of these jewels in the water, and the same ceremony was performed by each of the other priests. Whatever was done by the old priest was also done by the others in succession. Then the maiden brought kernels of corn on a tray, and these were in like manner placed on the water. She then placed a little brush near each of the priests. These brushes were made of the feathers of the beautiful warblers and humming-birds found in that region. Then she placed a tray of meal near each of the priests, and a tray of white sand, and a tray of red sand, and a tray of black sand. She then took from the niche in the wall a little stone vessel, in which had been ground some dried leaves, and placed it in the center of the space between the men. Then on a little willow-ware tray, woven of many-colored strands, she brought four pipes of the ancient pattern—hollow cones, in the apex of which were inserted the stems. Each of the priests filled his pipe with the ground leaves from the stone vessel. The maiden lighted a small fantastically painted stick and gave it to the priest, who lighted his pipe and smoked it with great vigor, swallowing the smoke until it appeared that his stomach and mouth were distended. Then, kneeling over the vase, he poured the smoke from his mouth into it, until it was filled, and the smoke piled over and gradually rose above him, forming a cloud.

Then the old man, taking one of the little feather brushes, dipped it into the vase of water and sprinkled the floor of the kiva, and, standing up, clasped his hands, turned his face upward, and prayed. "Mu-ing-wa! very good; thou dost love us, for thou didst bring us up from the lower world. Thou didst teach our fathers, and their wisdom has descended to us. We eat no stolen bread. No stolen sheep are found in our flocks. Our young men ride not the stolen ass. We beseech thee, Mu-ing-wa, that thou wouldst dip thy brush, made of the feathers of the birds of heaven, into the lakes of the skies, and scatter water over the earth, even as I scatter water over the floor of this kiva; Mu-ing-wa, very good."

Then the white sand was scattered over the floor, and the old man prayed that during the coming season Mu-ing-wa would break the ice in the lakes of heaven, and grind it into ice dust (snow) and scatter it over the land, so that during the coming winter the ground might be prepared for the planting of another crop. Then, after another ceremony with kernels of corn, he prayed that the corn might be impregnated with the life of the water, and made to bring forth an abundant harvest. After a ceremony with the jewels, he prayed that the corn might ripen, and that each kernel might be as hard as one of the jewels. Then this part of the ceremony ceased. The vases, and pitchers, and jewels, and other paraphernalia of the ceremony were placed away in the niche by the mother.

At day-break on the second morning, when the ceremonies had ceased, twenty five or thirty maidens came down into the kiva, disrobed themselves, and were re-clothed in gala dress, variously decorated with feathers and bells, each assisting the other. Then their faces were painted by the men in this wise: A man would take some paint in his mouth, thoroughly mix it with saliva, and with his finger paint the girl's face with one color, in such manner as seemed right to him, and she was then turned over to another man who had another color prepared. In this way their faces were painted yellow, red, and blue. When all was ready, a line was formed in the kiva, at the head of which was the grandmother, and at the foot the virgin priestess, who had attended through the entire ceremony. As soon as the line was formed below, the men, with myself, having in the meantime reclothed ourselves, went up into the court and were stationed on the top of the house nearest the entrance to the kiva. We found all the people of this village, and what seemed to me all the people of the surrounding villages, assembled on top of the houses, men, women, and children, all standing expectant.

As the procession emerged from the kiva by the ladder, the old woman commenced to chant. Slowly the procession marched about the court and around two or three times, and then to the center, where the maidens formed a circle, the young virgin priestess standing in the center. She held in her hand a beautifully wrought willow-work tray, and all the young men stood on the brink of the wall next to the plaza, as if awaiting a signal. Then the maiden, with eyes bandaged,[24] turned round and round, chanting something which I could not understand, until she should be thoroughly confused as to the direction in which the young men stood. Then she threw out of the circle in which she stood the tray which she held, and, at that instant, every young athlete sprang from the wall and rushed toward the tray and entered into the general conflict to see who should obtain it. No blows were given, but they caught each other about the waist and around the neck, tumbling and rolling about into the court until, at last, one got the tray into his possession for an instant, threw it aloft and was declared the winner. With great pride he carried it away. Then the women returned to the kiva. In a few minutes afterward they emerged again, another woman carrying a tray, and so the contests were kept up until each maiden had thrown a tray into the court-yard, and it had been won by some of the athletes. About ten o'clock these contests ended, and the people retired to their homes, each family in the village inviting its friends from the sur-

---

[24]Blindfolded.

rounding villages, and for an hour there was feasting and revelry. During the afternoon there were races, and afterward dancing, which was continued until midnight.

In a former article I have briefly described the system of picture-writings found in use among these people. These are rude etchings on the rocks or paintings on tablets of wood. They are simply mnemonic, and are, of course, without dates. A great buffalo hunt is recorded with a picture of a man standing in front of and pointing an arrow at one of these animals. The record of a great journey is made with a rude map. On the Cliff near Oraibi, I found a record like this etched on a stone. Below and to the left were three Spaniards, the leader with a sword, the two followers carrying spears. Above and to the right were three natives in an attitude of rolling rocks. Near by was a Spaniard prone on the ground, with a native pouring water on his head. Tal-ti, whose name means "peep of day," because he was born at dawn, explained to me that the record was made by their ancestors a very long time ago, and that the explanation had been handed down as follows: Their town was attacked by the Spaniards; the commander was a gallant fellow, who attempted to lead his men up the stone stairway to the town, but the besieged drove them back with rolling stones, and the Spanish captain was wounded and left by his followers. The people, in admiration of his valor, took him to a spring near by, poured water on him, dressed his wounds, and, when they were healed, permitted him to return.

Tal-ti's description of the scene was quite vivid, and even dramatic, especially when he described the charge of the Spaniards rushing forward and shouting their war cries, *"Santiago! Santiago! Santiago!"*

Thus in this desert land we find an agricultural people; a people living in stone houses, with walls laid in mortar and plastered within, houses two, three, four, five, or six stories high; a people having skill in the manufacture and ornamentation of pottery, raising cotton, and weaving and dyeing their own clothing, skilled in a system of picture-writings, having a vast store of mythology, and an elaborate, ceremonious religion; without beasts of burden, and having no knowledge of metals, all their tools being made of bones, stone, or wood. Such was their condition when found by the first

Europeans who invaded their lands. Early in the recorded history of this country they obtained from the Spaniards a few tools of iron, some sheep, which they raised for their flesh as well as for their wool, and asses, which they use as a means of transportation.

The seven hamlets of this province form only one of many groups discovered by those early Spanish adventurers. Altogether, about sixty towns were found by them; about half of these were destroyed, and, in all the remaining towns, except the seven, a new religion was imposed upon the people. It should rather be said that Christian forms and Christian ideas were ingrafted on the old pagan stock. Most of the towns outside of this province are watched over by Catholic priests, and the pagan rites and ceremonies are prohibited.[25] But occasionally the people steal away from their homes and assemble on the mountains or join the people of the "Province of Tusayan" in the kivas, and celebrate the rites of their ancient religion.

"Who are these people?" is a question often asked. Are they a remnant of some ancient invading race from the Eastern Continent? I think not. Linguistic evidence shows them to be nearly related to some of the nomadic tribes of the Rocky Mountains, such as the Shoshones, Utes, Pai Utes, and Comanches.[26] The region of country between the Rocky Mountains and the Sierras, stretching from northern Oregon to the Gulf of California, is occupied by many tribes speaking languages akin to one another. These town-building people seem to be a branch of this great family now, but a remnant of this branch is left;[27] but there was a time when they were a vast people. The ruins of these towns are found in great profusion throughout Nevada, Utah, Colorado, New Mexico, Arizona, and Southern California.

---

[25] In Powell's time, traditional pueblo and "ingrafted" Christian practices coexisted more harmoniously than he suggests.

[26] The languages of these tribes derive from a common Shoshonean source. Powell was the first to apprehend their common heritage.

[27] The phrase should probably read, "but a remnant of which branch is left." Perhaps a copy error was made in the original.

On every stream, and at almost every spring of importance, vestiges of this race may be found. Where Salt Lake City now stands, in that ancient time there stood a settlement of the people calling themselves Shi-nu-mos, a word signifying "We, the wise." I have visited nearly every settlement in the Territory of Utah, and many in the State of Nevada, and have never failed, on examination, to find evidences of an ancient town on the same site, or one near by. On the eastern slope of the Rocky Mountains they have also been found; one near Golden City by Captain Berthoud,[28] and many others on the same slope to the southward. I have found them on the western slope of the same system of mountains, on the Yampa, White and Grand Rivers; and Dr. Newberry and Mr. Jackson have found them in great abundance on the San Juan and its tributaries.[29] The history of the exploration of New Mexico and Arizona is replete with accounts of these vestiges of ancient life.

Over all this vast territory, in every beautiful valley and glen, by every stream of water and every spring, on the high mountains, on the cliffs, away out in the deserts of drifting sand, and down in the deep cañon gorges by which much of the country is traversed—everywhere are found ruins, stone implements or fragments of pottery.

How have these people been so nearly destroyed? From a somewhat careful examination of the facts at hand, I have an explanation to offer, though I cannot here give the fragments of evidence on which it rests. There are two great bodies of Indians in this country who are intruders—

---

[28]Probably E. L. Berthoud of the 2nd Colorado Infantry, namesake of 11,315-foot Berthoud Pass near Winter Park.

[29]John Strong Newberry, a geologist, visited Aztec Ruin on the Animas River, a tributary of the San Juan River, in 1859 and later described it (Elliott, *Great Excavations,* 55). William Henry Jackson, an accomplished photographer, led a small unit of the Hayden Survey into the watershed of the Rio Mancos, another tributary of the San Juan, in 1874. The Mancos drains the country around Mesa Verde, and although Jackson saw and photographed many ancestral puebloan sites, he failed to discover Mesa Verde's spectacular cliff dwellings. Not until 1888, when cowboys Bob Wetherill and Charles Mason searched one of the mesa's side canyons for lost stock, did the famous Cliff Palace become known to the world (Goetzmann, *Exploration and Empire,* 523–25).

the Navajos and Apaches, and a number of small tribes in California[30] who speak Athabascan languages, and who originally dwelt far to the north in British America. The Pueblo people call them their northern enemies.[31] It seems that these people gradually spread to the south, attracted perhaps by the wealth accumulated by an agricultural and economic people; and, as they swept southward, from time to time, in bold excursions, town after town, and hamlet after hamlet was destroyed; the people were driven into the cañons and among the cliffs, and on the advent of the white man to this continent, only the sixty towns which I have mentioned remained. Of these, there are now but thirty. Of the former inhabitants of the thirty destroyed since the first invasion of the country by the Spaniards, some, at least, have become nomadic for the Co-a-ni-nis and Wal-la-pais, who now live in the rocks and deep gorges of the San Francisco Plateau, claim that at one time they dwelt in pueblos, near where Zunia now stands.[32]

Interested as we were in this strange people, time passed rapidly, and our visit among them was all too short; but, at last, the time came for us to leave. When we were ready to start we were joined by a small delegation of the Indians, who proposed to travel with us for a few days. We made our way to Fort Defiance, thence to Fort Wingate,[33] and still on to the East until we reached the Valley of the Rio Grande del Norte. Here we stopped for a day to visit the ancient town of Jemez, and then proceeded to Santa Fe, where our long journey on horseback ended.

---

[30]California Athapascans include the Cahto, Chilula, Hupa, Lassik, Mattole, Nongatl, Sinkyone, Tolowa, Wailaki, and Whilkut, all of whose territories lie in the northwest corner of the state.

[31]Utes may also be added to the list.

[32]*Coanini, Coconini,* and a range of cognates were once used to refer to upland Yuma-speaking natives, especially the Havasupai and Walapai, whose ancestral territories include the plateaus south and west of the Grand Canyon. By *Zunia* Powell refers to Zuni.

[33]Near Gallup, New Mexico.

———◆———

PART IV

Report on the Lands
of the Arid Region

———◆———

I N 1954 BERNARD DEVOTO, a Utah native, national gadfly, and author of a trio of prize-winning histories of the nineteenth-century West, described John Wesley Powell's *Report on the Lands of the Arid Region* as

One of the most remarkable books ever written by an American. In the whole range of American experience from Jamestown on there is no book more prophetic. It is a scientific prophecy and it has been fulfilled—experimentally proved. Unhappily the experimental proof has consisted of human and social failure and the destruction of land. It is a document as basic as *The Federalist* but it is a tragic document. For it was published in 1878 and if we could have acted on it in full, incalculable loss would have been prevented and the United States would be happier and wealthier than it is. We did not even make an effective effort to act on it till 1902. . . . We are still far short of catching up with it. The twist of the knife is that meanwhile irreversible actions went on out west and what we did in error will forever prevent us from catching up with it altogether.[1]

---

[1] From DeVoto's introduction to Stegner's *Beyond the Hundredth Meridian,* xxii. DeVoto's histories include *The Year of Decision: 1846, Across the Wide Missouri,* and *The Course of Empire.*

DeVoto was never one to mince words, but if he was guilty of overstate-
ment in his appraisal of Powell's *Report on the Land of the Arid Region,* he
was nevertheless fundamentally right about its significance. In its quiet
way—beginning, for instance, with a long disquisition on, of all things, the
weather—the report is Promethean in scope, profound in implication. In it,
Powell laid out a plan for the settlement of the entire West.

No one had attempted such a thing before. For one thing, the region,
which sprawled from the Canadian to the Mexican border and from the
grassland sea of the Great Plains to the Pacific coast, was so vast and varied
that it defied synthetic understanding. Powell, however, had personally
logged thousands of western miles by river, on foot, and by horseback,
wagon, and train, and he seems to have been attentive and observant every
yard of the way. For another, few people saw the need to treat the West differently from the rest of the nation: It was simply more of America, and it
was assumed that settlement would proceed there as it had proceeded in
the East. But not in Powell's view. He saw that the West was distinguished
by aridity, with the result that eastern ways and institutions would not work
there. For yet another, no one had previously attempted to synthesize the
enormous body of information that was rapidly accumulating about the
West. It derived from federally sponsored surveys, from the reports of military expeditions, from the personal testimony of settlers, traders, trappers,
and travelers, and from the pronouncements of the frontier newspapers
that seemed to spring into being wherever three or more tents were pitched
together. The task of synthesis was tailor-made for Powell, who brought to
it a wealth of personal experience and observation. It was the habit of his
mind always to fit new data into the pattern of the whole, to build continually the mosaic of a total view. The *Report* is his mosaic of the West, as he
conceived it in 1878.

The *Report on the Lands of the Arid Region* came into existence partly
as a result of the reorganization of the western surveys. Through most of
the 1870s, the federal government funded four different and competing
efforts to inventory and map western lands: Powell's own Geographical
and Geological Survey of the Rocky Mountain Region, Ferdinand V.

Hayden's United States Geological and Geographical Survey of the Territories, Clarence King's United States Geological and Geographical Survey of the Fortieth Parallel, and Lt. George M. Wheeler's Geographical Surveys West of the 100th Meridian.[2] Competition between the surveys was intense, both for funding and for lands to be mapped. In 1873 parties from the Wheeler and Hayden surveys clashed at South Park, Colorado Territory, each asserting its right to run lines through lands claimed by the other. Congress took note of the incident and made a few administrative adjustments, but the four rival surveys continued. One of the adjustments was to reassign Powell's survey, by far the smallest of the four in budgetary terms, from the sponsorship of the Smithsonian Institution to the Interior Department, where it ostensibly became a unit of the Hayden survey. In practical terms, however, the new arrangement did not subordinate Powell to Hayden, and the two surveys continued to operate independently.

Meaningful change had to wait until the election of Rutherford B. Hayes in 1876, when the Grant era ended and reformers took the reins of government. Carl Schurz became secretary of interior and soon insisted that his department's two surveys not duplicate their efforts. Allowed to choose between geology and ethnology, Hayden elected to focus his efforts on geology, leaving the other field for Powell. Powell dutifully honored the agreement by suspending the initiation of new geological work, but rather than turn immediately to ethnology, he seized the opportunity to educate the government and the populace about the physical realities of the western territories. He devoted his own energies and assigned his best men, who included Clarence Dutton and Grove Karl Gilbert, to a penetrating geographic study of nearly the entire public domain. Powell intended to analyze the potential of the land to support human settlement and then to eval-

---

[2] Two points of reference: The 100th meridian marks the east boundary of the Texas panhandle and roughly bisects the plains states of Kansas, Nebraska, and the two Dakotas; the 40th parallel divides Kansas from Nebraska and runs through northern portions of Colorado, Utah, Nevada, and California.

uate the land laws of the United States in terms of their effectiveness in releasing that potential.

In previous travels, Powell had closely observed the irrigation practices of Mormons, Pueblo Indians, and Hispanic New Mexicans, and he concluded that the success of continued settlement would similarly depend on a foundation of oasis agriculture. Late in life, Powell would look back and say that his interest in the subject was awakened during his first western expedition in 1867 to the mountains of Colorado:

> About the campfire on various occasions the problem of the future of the arid region was discussed. The central opinion expressed was that it would always be dependent upon mining industries, and that mining must necessarily, to some extent, be precarious. I took the ground that ultimately agriculture and manufacturing would be developed on a large scale, and to substantiate my opinion in relation to agriculture, I called attention to the vast agricultural resources of the arid regions of Egypt, Persia, India, China, and other countries of the world where agriculture was dependent upon irrigation, and I affirmed that in a very few decades all the waters of the arid region of the United States would be used in irrigation for agricultural purposes. During these conversations I became deeply interested in the problem.[3]

Powell may have exaggerated both his prescience and his familiarity with the history of ancient civilizations, such as it existed in 1867, but there can be little doubt that he came early to the question of agriculture in the arid lands and pondered it long and hard. One conclusion fairly leapt from the facts: Eastern institutions were ill-adapted to western conditions. The land laws and policies of the United States and the customs of its people had developed in a rain-blessed climate and were predicated on an individualistic model of human endeavor, which that climate made possible. As Powell well knew, a family working alone could carve a farmstead and eke out a liv-

---

[3] As quoted by Pisani, *To Reclaim a Divided West,* 144. This statement comes from an undated document in the National Archives, which Donald Worster, Powell's most recent biographer, places in 1890.

ing from the eastern forests and tallgrass prairies. For 20 restless years, his own family, as peripatetic as any other Americans, had done just that in western New York, southeastern Ohio, Wisconsin, and Illinois. In the arid West, however, the lands that might allow a family to make it on their own were limited to valley bottoms that small streams might irrigate. The diversion of larger streams, on which depended the greater part of the region's potential to support settlement, would be delayed "until coöperative labor can be organized or capital induced to assist." It was clear to Powell that without reform, the nation's land laws, which enshrined the 160-acre homestead as a cultural ideal, would only obstruct the needed cooperative efforts. Worse, they would mislead countless migrants to the West into believing that the quarter-section homestead, an artifact of the humid East, might afford them enough of a living to justify their sweat and labor. In fact, what the land laws mainly guaranteed was suffering for individual families and burgeoning opportunities in real estate speculation for entrepreneurs savvy enough to circumvent the laws. Quite rightly, Powell viewed the Homestead Act and related measures as a hoax on aspiring westerners, and his report offered a blueprint for the reforms that would produce the more appropriate organizational structure required by a communitarian West.

To be sure, the report reflected Powell's personal ambition. It called for the inventory and classification of western lands on a grand scale, a task Powell was hungry to undertake. The report was also a work of extreme idealism, and not just in its appeal to communitarian values. In it Powell laid out his plan for what today might be called the sustainable development of 40 percent of the continental United States. Some of his ideas were untried; many ran counter to people's expectations and conventional behavior. With the notable exception of Utah's Mormons, there was little evidence that westering Euro-Americans would accept Powell's prescription for cooperative settlement. But the report was no less a work of pragmatism: It was built on a foundation of empirical observation (including the deliberately methodical opening discussion of rainfall), and it culminated in recommendations for specific, tangible legislative action.

Here are all of Powell's apparent contradictions in one package. Not only does the *Report* wed his pragmatism to a utopian vision; it also joins Powell's faith in science to his humanism. As a scientist, he fervently believed that experts must execute the work of survey and land classification on which his plan depended. Until the technical men—his men—accomplished this fundamentally scientific task, he believed that the settlement of the West should be reined in.

But he did not believe that the technical men should remain in control, and critics who derogate Powell as an apostle of bureaucracy often miss this point. Once the framework for settlement was in place, Powell believed that the technical men—and the federal government they represented—should step aside, allowing settlers through their cooperative communities to assume control of the resources on which they depended. This hybrid position, taken entire, represents one of Powell's most original and least popular ideas. Westerners applauded his endorsement of local control but chafed at his insistence on methodical survey and classification. They wanted action, not study. Leaders of the emergent conservation movement, such as Bernard Fernow, Gifford Pinchot, and Charles Sprague Sargent, shared his faith in science and in the importance of entrusting resource allocation to impartial experts, but they were aghast that he would assign the actual ownership and management of those resources to the untrained masses. They had seen too much market hunting and too much rape-and-run timber cutting to have confidence in the self-restraint of local interests. In their view, the protection of resources depended on control and regulation by the highest possible level of government.

To the end of his days, Powell maintained his lonely position as a scientific populist, elaborating it in interesting ways. In 1890, he published a series of articles in *Century Magazine* that revisit the problems addressed in the *Arid Lands* report (see Part VI). In them he argues again for local control but with retention of federal ownership of all but the irrigable lands. He asserts in those articles that local interests can be expected to manage their resources responsibly, not because of their presumed goodness but because of their ability to recognize the interdependence of the resources and their interests in them. Farmers, he reasoned, would not abuse the watersheds on which they depended for water, and if they did, they would quickly learn the consequences of their actions. The government should allow them independence and self-determination, its message to them being, "With wisdom you may prosper, but with folly you must fail."

Powell's position did not prevail, and more than a century later, many westerners still fight with the federal government over local control of forest and range resources. But Powell's ideas keep resurfacing. Currently some of the most radical experiments in western land management essentially embody Powell's recommendation for federal ownership and local control of western watersheds.[4] His proposals for organizing irrigation districts and administering grazing leases have also taken root, although in the latter case only after a debacle of overgrazing on the public domain that lasted more than half a century.

Not all the ideas contained in the *Arid Lands* report were original with Powell, but no one else before or since has succeeded so well in integrating and presenting them as a single interlocking whole. Powell was part Cassandra, part Pollyanna. He held that the course of western development in

---

[4] In 2000, for instance, the federal government purchased the 89,000 acres of the Baca Ranch in New Mexico, assigning its management not to a federal agency but to a trust directed by a board of individuals drawn mainly from the surrounding area. Explicitly, the members of the board were to possess diverse skills in various areas of land management (such as forestry, grazing, wildlife management, and conservation); implicitly, they were to represent a range of political points of view, thereby reflecting the diversity of the multiple potential constituencies of the Baca Ranch. Other recent land management experiments emphasizing broad local participation, each producing thoroughly different results, include the work of the Quincy Library Group in California, the Malpai Borderlands Group in southwestern New Mexico and southeastern Arizona, and various projects of the Greater Yellowstone Coalition and the Grand Canyon Trust, to name only a few. Dan Kemmis, a former mayor of Missoula and head of the University of Montana's Center for the Rocky Mountain West, provides an intellectual framework for these kinds of approaches in *Community and the Politics of Place,* and Michael Vincent McGinnis has collected essays that explore the topic from a range of perspectives in the anthology *Bioregionalism.* It should be noted that the term *bioregionalism,* under which most of these initiatives might be placed, means different things to different people. The ranchers of the Malpai Borderlands Group, for instance, might not readily identify themselves with the bioregional vision described by Kirkpatrick Sale in *Dwellers in the Land,* which propounds a political and economic philosophy emphasizing extreme regional self-sufficiency and nonmaterialism.

1878 was wrong and that it would lead to misuse of resources and enormous human suffering by encouraging people "to establish homes where they cannot maintain themselves."[5] Since then, the facts have borne out his predictions, never more horrifically than in the harsh days of the Dust Bowl, and no student of western history can today argue that his warnings were misguided. But Powell also presented an alternative. It was a vision of cooperative, small farming and ranching communities, with little concentration of ownership, providing an agrarian foundation for a western society that he acknowledged would also develop a range of mining and manufacturing industries. Assuredly, it was too good to be true, and the reality of its implementation would certainly have fallen short of the plan's abstract perfection. Powell's plan for acquiring large grazing homesteads, for instance, probably would have lent itself to abuse as much as any previous plan for the settlement and sale of public domain. Nevertheless, Powell's alternative might well have been preferable to what actually transpired. We will never know. The path he offered will forever remain the road not taken of western American history. The people whom it was to serve rejected it because they felt that it imposed on them too many hardships in the form of delay and regulation. Even more profoundly, it contradicted their belief, however exaggerated or misplaced it may have been, in their own independence and individualism. The homestead ideal lay at the heart of American national mythology, and it would not soon be dislodged.

Still, it is no mere coincidence that the course of actual events in the disposition and management of western lands has veered again and again in the direction Powell advised. His vision may not have been perfect, but no one in his day saw farther or more clearly into the future than he. We may hope in vain that we learn to see the future of our own time with equal clarity.

On April 1, 1878, Powell delivered the *Report* to J. A. Williamson, the commissioner of the General Land Office, who dutifully transmitted it to his superior, Carl Schurz, the secretary of the interior. On April 3, Schurz

---

[5] As quoted by Rabbitt, "John Wesley Powell: Pioneer Statesman of Federal Science," 18.

sent the document to Samuel J. Randall, the speaker of the House of Representatives. Without openly endorsing Powell's conclusions, Schurz wrote, "In view of the importance of rendering the vast extent of country referred to available for agricultural and grazing purposes, I have the honor to commend the views set forth by Major Powell and the bills submitted herewith to the consideration of Congress." Randall referred the report to the Appropriations Committee, which ordered 1,800 copies printed. That stock was soon exhausted, and in March of the following year, the Senate and House jointly ordered a second printing of 5,000 copies: 1,000 for the Senate and 2,000 each for the House and the Department of Interior.

The following selections from the report include Powell's preface and the first two chapters, which express his overall assessment of the region and his recommendations for land law reform. The rest of the *Report* becomes increasingly technical and detailed and includes the work of several additional authors. The omitted sections are by no means without interest, but they are not essential for appreciating Powell's vision for the arid lands. Nevertheless, the table of contents of the report is reprinted here, after Powell's preface, in order to give some idea of the entire report's breadth and depth.

Preface and Table of Contents

Preface

It was my intention to write a work on the Public Domain. The object of the volume was to give the extent and character of the lands yet belonging to the Government of the United States. Compared with the whole extent of these lands, but a very small fraction is immediately available for agriculture; in general, they require drainage or irrigation for their redemption.

It is true that in the Southern States there are some millions of acres, chiefly timber lands, which at no remote time will be occupied for agricultural purposes. Westward toward the Great Plains, the lands in what I have, in the body of this volume, termed the Humid Region have passed from the hands of the General Government. To this statement there are some small exceptions here and there—fractional tracts, which, for special reasons, have not been considered desirable by persons in search of lands for purposes of investment or occupation.

In the Sub-Humid Region settlements are rapidly extending westward to the verge of the country where agriculture is possible without irrigation.

In the Humid Region of the Columbia the agricultural lands are largely covered by great forests, and for this reason settlements will progress slowly, as the lands must be cleared of their timber.

The redemption of the Arid Region involves engineering problems requiring for their solution the greatest skill. In the present volume only these lands are considered. Had I been able to execute the original plan to my satisfaction, I should have treated the coast swamps of the South Atlantic and the Gulf slopes, the Everglade lands of the Floridian peninsula, the flood plain lands of the great rivers of the south, which have heretofore been made available only to a limited extent by a system of levees, and the lake swamp lands found about the headwaters of the Mississippi and the region of the upper Great Lakes.[1] All of these lands require either drainage or protection from overflow, and the engineering problems involved are of diverse nature. These lands are to be redeemed from excessive humidity, while the former are to be redeemed from excessive aridity. When the excessively humid lands are redeemed, their fertility is almost inexhaustible, and the agricultural capacity of the United States will eventually be largely increased by the rescue of these lands from their present valueless condition. In like manner, on the other hand, the arid lands, so far as they can be redeemed by irrigation, will perennially yield bountiful crops, as the means for their redemption involves their constant fertilization.[2]

To a great extent, the redemption of all these lands will require extensive and comprehensive plans, for the execution of which aggregated capital or coöperative labor will be necessary. Here, individual farmers, being poor men, cannot undertake the task. For its accomplishment a wise prevision, embodied in carefully considered legislation, is necessary. It was my purpose not only to consider the character of the lands themselves, but also the

---

[1] The intensity of Powell's ambition to create a nationwide system of land classification is fully evident, as is his view, typical of his day, that swamps and deserts should and eventually would be "redeemed." It is worth remarking that land reclamation, as discussed in Powell's day, borrowed heavily from the vocabulary of Mormon and Protestant religion, and the assumptions implicit in such borrowings suggest a kind of evangelical mission. To say that land might be redeemed implies that it has fallen from a more perfect state and that to reclaim it would contribute to the accomplishment of a greater destiny.

[2] Powell refers to the idea that irrigation water delivers nutrients to the land it waters, as the annual floods of the Nile formerly (before the Aswan High Dam) nourished the farms of Egypt with a layer of silt. In actual practice, silty irrigation water tends to clog the canals and ditches that distribute it, and the promise of fertilization becomes instead an unwanted mechanical duty of dredging and cleaning. Some irrigation systems use settling ponds to avoid the problem.

engineering problems involved in their redemption, and further to make suggestions for the legislative action necessary to inaugurate the enterprises by which these lands may eventually be rescued from their present worthless state. When I addressed myself to the broader task as indicated above, I found that my facts in relation to some of the classes of lands mentioned, especially the coast swamps of the Gulf and some of the flood plain lands of the southern rivers, were too meager for anything more than general statements. There seemed to be no immediate necessity for discussion of these subjects; but to the Arid Region of the west thousands of persons are annually repairing, and the questions relating to the utilization of these lands are of present importance. Under these considerations I have decided to publish that portion of the volume relating to the arid lands, and to postpone to some future time that part relating to the excessively humid lands.

In the preparation of the contemplated volume I desired to give a historical sketch of the legislation relating to swamp lands and executive action thereunder; another chapter on bounty lands and land grants for agricultural schools, and still another on land grants in aid of internal improvements—chiefly railroads. The latter chapter has already been prepared by Mr. Willis Drummond, Jr., and as the necessary map is ready I have concluded to publish it now, more especially as the granted lands largely lie in the Arid Region.[3] Mr. Drummond's chapter has been carefully prepared and finely written, and contains much valuable information.

To the late Prof. Joseph Henry, secretary of the Smithsonian Institution,[4] I am greatly indebted for access to the records of the Institution relating to rainfall. Since beginning my explorations and surveys in the far west, I have received the counsel and assistance of the venerable professor on all important matters relating to my investigations; and whatever of value has been accomplished is due in no small part to his wisdom and advice. I can-

---

[3] The referenced map, not included in this volume, was titled "Map of the United States exhibiting the grants of lands made by the General Government to aid in the construction of Railroads and Wagon Roads."

[4] Powell named the last discovered mountain range in the coterminous United States—Utah's Henry Mountains—for this beloved and distinguished patron.

not but express profound sorrow at the loss of a counselor so wise, so patient, and so courteous.

I am also indebted to Mr. Charles A. Schott, of the United States Coast Survey, to whom the discussion of the rain gauge records has been intrusted by the Smithsonian Institution, for furnishing to me the required data in advance of publication by himself.

Unfortunately, the chapters written by Messrs. Gilbert, Dutton, Thompson, and Drummond[5] have not been proof-read by themselves, by reason of their absence during the time when the volume was going through the press; but this is the less to be regretted from the fact that the whole volume has been proof-read by Mr. J. C. Pilling, whose critical skill is all that could be desired.

Table of Contents[6]

---

[5]Gilbert, Dutton, and Thompson were among Powell's most outstanding co-workers. Grove Karl Gilbert (1843–1918), a geologist with the Wheeler survey, met Powell in 1872 when the latter was fresh from his exploration of the Colorado. Two years later Gilbert jumped ship and joined Powell's survey, staying with it and the combined U.S. Geological Survey until his death. In one of his greatest insights, Gilbert divined the Pleistocene existence of Utah's extinct Lake Bonneville. He also coined the term *Colorado Plateau*. Clarence E. Dutton (1841–1912), another brilliant geologist, was especially gifted as a scientific writer. His *Tertiary History of the Grand Canyon District* (1882) is a descriptive and explanatory masterwork. Almon Thompson (1839–1906), Powell's brother-in-law, was a member of the 1871–72 Colorado River expedition and, although self-taught, became an accomplished topographer and mapmaker, remaining with the U.S. Geological Survey until 1904.

[6]Only chapters 1 and 2 are included in the present volume, but the Table of Contents of the *Report on the Arid Lands* gives some idea of its comprehensiveness and the weight of factual evidence provided to buttress its arguments.

Physical Characteristics of the Arid Region

The eastern portion of the United States is supplied with abundant rainfall for agricultural purposes, receiving the necessary amount from the evaporation of the Atlantic Ocean and the Gulf of Mexico; but westward the amount of aqueous precipitation diminishes in a general way until at last a region is reached where the climate is so arid that agriculture is not successful without irrigation.[1] This Arid Region begins about midway in the Great Plains and extends across the Rocky Mountains to the Pacific Ocean. But on the northwest coast there is a region of greater precipitation, embracing western Washington and Oregon and the northwest corner of California. The winds impinging on this region are freighted with moisture derived from the great Pacific currents; and where this water-laden atmosphere strikes the western coast in full force, the precipitation is excessive, reaching a maximum north of the Columbia River of 80 inches annually. But the rainfall rapidly decreases from the Pacific Ocean eastward to the summit of the Cascade Mountains. It will be convenient to designate this humid area as the Lower Columbia Region. Rain gauge records have not been made to such an extent as to enable us to define its eastern and southern boundaries, but as they are chiefly along high mountains, definite boundary lines are unimportant in the consideration of agricultural resources and the questions relating thereto. In like manner on the

---

[1] Advice for the uncertain reader: Do not be discouraged that Powell begins this report with a tedious review of rainfall information. He is laying a foundation and does not choose to attract attention, by way of narrative flourish, to the mere pouring of concrete. The revolutionary arguments he is about to make require a strong footing. (Incidentally, his meteorology ignores the contribution of Pacific storms to eastern weather.)

east the rain gauge records, though more full, do not give all the facts necessary to a thorough discussion of the subject; yet the records are such as to indicate approximately the boundary between the Arid Region, where irrigation is necessary to agriculture, and the Humid Region, where the lands receive enough moisture from the clouds for the maturing of crops. Experience teaches that it is not wise to depend upon rainfall where the amount is less than 20 inches annually, if this amount is somewhat evenly distributed throughout the year; but if the rainfall is unevenly distributed, so that "rainy seasons" are produced, the question whether agriculture is possible without irrigation depends upon the time of the "rainy season" and the amount of its rainfall. Any unequal distribution of rain through the year, though the inequality be so slight as not to produce "rainy seasons," affects agriculture either favorably or unfavorably. If the spring and summer precipitation exceeds that of the fall and winter, a smaller amount of annual rain may be sufficient; but if the rainfall during the season of growing crops is less than the average of the same length of time during the remainder of the year, a greater amount of annual precipitation is necessary. In some localities in the western portion of the United States this unequal distribution of rainfall through the seasons affects agriculture favorably, and this is true immediately west of the northern portion of the line of 20 inches of rainfall, which extends along the plains from our northern to our southern boundary.

The isohyetal[2] or mean annual rainfall line of 20 inches, as indicated on the rain chart accompanying this report, begins on the southern boundary of the United States, about 60 miles west of Brownsville, on the Rio Grande del Norte, and intersects the northern boundary about 50 miles east of Pembina.[3] Between these two points the line is very irregular, but in middle latitudes makes a general curve to the westward. On the southern portion of the line the rainfall is somewhat evenly distributed through the seasons, but

---

[2] An isohyet is a line drawn on a map connecting points receiving equal rainfall.
[3] Pembina lies in the extreme northeast corner of North Dakota, opposite the Canadian and Minnesota borders.

along the northern portion the rainfall of spring and summer is greater than that of fall and winter, and hence the boundary of what has been called the Arid Region runs farther to the west. Again, there is another modifying condition, namely, that of temperature. Where the temperature is greater, more rainfall is needed; where the temperature is less, agriculture is successful with a smaller amount of precipitation. But geographically this temperature is dependent upon two conditions—altitude and latitude. Along the northern portion of the line latitude is an important factor, and the line of possible agriculture without irrigation is carried still farther westward. This conclusion, based upon the consideration of rainfall and latitude, accords with the experience of the farmers of the region, for it is a well known fact that agriculture without irrigation is successfully carried on in the valley of the Red River of the North, and also in the southeastern portion of Dakota Territory. A much more extended series of rain-gauge records than we now have is necessary before this line constituting the eastern boundary of the Arid Region can be well defined. It is doubtless more or less meandering in its course throughout its whole extent from south to north, being affected by local conditions of rainfall, as well as by the general conditions above mentioned; but in a general way it may be represented by the one hundredth meridian, in some places passing to the east, in others to the west, but in the main to the east.

The limit of successful agriculture without irrigation has been set at 20 inches, that the extent of the Arid Region should by no means be exaggerated; but at 20 inches agriculture will not be uniformly successful from season to season. Many droughts will occur; many seasons in a long series will be fruitless; and it may be doubted whether, on the whole, agriculture will prove remunerative. On this point it is impossible to speak with certainty. A larger experience than the history of agriculture in the western portion of the United States affords is necessary to a final determination of the question.[4]

In fact, a broad belt separates the Arid Region of the west from the Humid Region of the east. Extending from the one hundredth meridian

_____

4An admirable admission of the insufficiency of available data.

eastward to about the isohyetal line of 28 inches, the district of country thus embraced will be subject more or less to disastrous droughts, the frequency of which will diminish from west to east. For convenience let this be called the Sub-Humid Region.[5] Its western boundary is the line already defined as running irregularly along the one hundredth meridian. Its eastern boundary passes west of the isohyetal line of 28 inches of rainfall in Minnesota, running approximately parallel to the western boundary line above described. Nearly one-tenth of the whole area of the United States, exclusive of Alaska, is embraced in this Sub-Humid Region. In the western portion disastrous droughts will be frequent; in the eastern portion infrequent. In the western portion agriculturists will early resort to irrigation to secure immunity from such disasters, and this event will be hastened because irrigation when properly conducted is a perennial source of fertilization, and is even remunerative for this purpose alone; and for the same reason the inhabitants of the eastern part will gradually develop irrigating methods.[6] It may be confidently expected that at a time not far distant irrigation will be practiced to a greater or lesser extent throughout this Sub-Humid Region.[7] Its settlement presents problems differing materially from those pertaining to the region to the westward. Irrigation is not immediately necessary, and hence agriculture does not immediately depend upon capital. The region may be settled and its agricultural capacities more or less developed, and the question of the construction of irrigating canals may be a matter of time and convenience. For many reasons, much of the sub-humid belt is attractive to settlers: it is almost destitute of forests, and for this reason is more readily subdued, as the land is ready for the plow. But because of the lack of forests the country is more dependent upon railroads for the transportation of building and fencing materials and for fuel. To a large extent it is a region

[5] See map 7, "Arid Region of the United States Showing Drainage Districts," for a delineation of the boundary between Powell's conception of the Sub-Humid Region and the Arid Region.

[6] See Selection 6, note 2.

[7] In fact, irrigation (especially from groundwater) is today extensive in many areas immediately east of the 100th meridian.

where timber may be successfully cultivated. As the rainfall is on a general average nearly sufficient for continuous successful agriculture, the amount of water to be supplied by irrigating canals will be comparatively small, so that its streams can serve proportionally larger areas than the streams of the Arid Region. In its first settlement the people will be favored by having lands easily subdued, but they will have to contend against a lack of timber. Eventually this will be a region of great agricultural wealth, as in general the soils are good. From our northern to our southern boundary no swamp lands are found, except to some slight extent in the northeastern portion, and it has no excessively hilly or mountainous districts. It is a beautiful prairie country throughout, lacking somewhat in rainfall; but this want can be easily supplied by utilizing the living streams; and, further, these streams will afford fertilizing materials of great value.

The Humid Region of the lower Columbia and the Sub-Humid Region of the Great Plains have been thus briefly indicated in order that the great Arid Region, which is the subject of this paper, may be more clearly defined.

The Arid Region

The Arid Region is the great Rocky Mountain Region of the United States, and it embraces something more than four-tenths of the whole country, excluding Alaska. In all this region the mean annual rainfall is insufficient for agriculture, but in certain seasons some localities, now here, now there, receive more than their average supply. Under such conditions crops will mature without irrigation. As such seasons are more or less infrequent even in the more favored localities, and as the agriculturist cannot determine in advance when such seasons may occur, the opportunities afforded by excessive rainfall cannot be improved.

In central and northern California an unequal distribution of rainfall through the seasons affects agricultural interests favorably. A "rainy season" is here found, and the chief precipitation occurs in the months of December–April. The climate, tempered by mild winds from the broad expanse of Pacific waters, is genial, and certain crops are raised by sowing the seeds immediately before or during the "rainy season," and the watering which they receive causes the grains to mature so that fairly remunerative crops are produced. But here again the lands are subject to the droughts of abnormal seasons. As many of these lands can be irrigated, the farmers of the country are resorting more and more to the streams, and soon all the living waters of this region will be brought into requisition.

In the tables of a subsequent chapter this will be called the San Francisco Region.

Again in eastern Washington and Oregon, and perhaps in northern Idaho, agriculture is practiced to a limited extent without irrigation. The conditions of climate by which this is rendered possible are not yet fully understood. The precipitation of moisture on the mountains is greater than on the lowlands, but the hills and mesas adjacent to the great masses of mountains receive a little of the supply condensed by the mountains themselves, and it will probably be found that limited localities in Montana, and even in Wyoming, will be favored by this condition to an extent sufficient to warrant agricultural operations independent of irrigation. These lands, however, are usually supplied with living streams, and their irrigation can be readily effected, and to secure greater certainty and greater yield of crops irrigation will be practiced in such places.

IRRIGABLE LANDS

Within the Arid Region only a small portion of the country is irrigable. These irrigable tracts are lowlands lying along the streams. On the mountains and high plateaus forests are found at elevations so great that frequent summer frosts forbid the cultivation of the soil. Here are the natural timber lands of the Arid Region—an upper region set apart by nature for the growth of timber necessary to the mining, manufacturing, and agricultural industries of the country. Between the low irrigable lands and the elevated forest lands there are valleys, mesas, hills, and mountain slopes bearing grasses of greater or less value for pasturage purposes.

Then, in discussing the lands of the Arid Region, three great classes are recognized—the irrigable lands below, the forest lands above, and the pasturage lands between. In order to set forth the characteristics of these lands and the conditions under which they can be most profitably utilized, it is deemed best to discuss first a somewhat limited region in detail as a fair type of the whole. The survey under the direction of the writer has been extended over the greater part of Utah, a small part of Wyoming and Colorado, the northern portion of Arizona, and a small part of Nevada, but it is proposed to take up for this discussion only the area embraced in Utah Territory.[8]

In Utah Territory agriculture is dependent upon irrigation. To this statement there are some small exceptions. In the more elevated regions there are tracts of meadow land from which small crops of hay can be taken: such lands being at higher altitudes need less moisture, and at the same time receive a greater amount of rainfall because of the altitude; but these meadows have been, often are, and in future will be, still more improved by irrigation. Again, on the belt of country lying between Great Salt Lake and the Wasatch Mountains the local rainfall is much greater than the general rainfall of the region. The water evaporated from the lake is carried by the westerly winds to the adjacent mountains on the east and again condensed, and the rainfall thus produced extends somewhat beyond the area occupied by the mountains, so that the foot hills and contiguous bench lands receive a modicum of this special supply. In some seasons this additional supply is enough to water the lands for remunerative agriculture, but the crops grown will usually be very small, and they will be subject to seasons of extreme drought, when all agriculture will result in failure. Most of these lands can be irrigated, and doubtless will be, from a consideration of the facts already stated, namely, that crops will thereby be greatly increased and immunity from drought secured. Perhaps other small tracts, on account of their subsoils, can be profitably cultivated in favorable seasons, but all of

---

[8] Powell's focus on Utah was in part determined by the intense competition among the great surveys of the West. Ferdinand V. Hayden was operating in the northern Rockies, Clarence King in California, and Lt. George M. Wheeler in the Southwest and Nevada. Utah, neglected by the others, offered opportunity for Powell and his men to operate freely.

these exceptions are small, and the fact remains that agriculture is there dependent upon irrigation. Only a small part of the territory, however, can be redeemed, as high, rugged mountains and elevated plateaus occupy much of its area, and these regions are so elevated that summer frosts forbid their occupation by the farmer. Thus thermic conditions limit agriculture to the lowlands, and here another limit is found in the supply of water. Some of the large streams run in deep gorges so far below the general surface of the country that they cannot be used; for example, the Colorado River runs through the southeastern portion of the Territory and carries a great volume of water, but no portion of it can be utilized within the Territory from the fact that its channel is so much below the adjacent lands. The Bear River,[9] in the northern part of the Territory, runs in a somewhat narrow valley, so that only a portion of its waters can be utilized. Generally the smaller streams can be wholly employed in agriculture, but the lands which might thus be reclaimed are of greater extent than the amount which the streams can serve;[10] hence in all such regions the extent of irrigable land is dependent upon the volume of water carried by the streams.

In order to determine the amount of irrigable land in Utah it was necessary to determine the areas to which the larger streams can be taken by proper engineering skill, and the amount which the smaller streams can serve. In the latter case it was necessary to determine first the amount of land which a given amount or unit of water would supply, and then the volume of water running in the streams; the product of these factors giving the extent of the irrigable lands. A continuous flow of one cubic foot of water per second was taken as the unit, and after careful consideration it was assumed that this unit of water will serve from 80 to 100 acres of land.[11] Usually the computations have been made on the basis of 100 acres. This unit was determined in

---

[9] The Bear River feeds Bear Lake, at the eastern extremity of the Utah–Idaho border

[10] Powell's emphasis on the limited extent of western land that might be reclaimed by irrigation will be a recurring point of tension for the next 20 years between him and other irrigation advocates who remained convinced of an unlimited future for irrigation.

[11] Measurements in cubic feet per second reflect the rate of flow. At the time of the report, the quantity of water delivered for irrigation was also generally denoted, albeit indirectly, by

the most practical way—from the experience of the farmers of Utah who have been practicing agriculture for the past thirty years. Many of the farmers will not admit that so great a tract can be cultivated by this unit. In the early history of irrigation in this country the lands were oversupplied with water, but experience has shown that irrigation is most successful when the least amount of water is used necessary to a vigorous growth of the crops; that is, a greater yield is obtained by avoiding both scanty and excessive watering; but the tendency to over-water the lands is corrected only by extended experience. A great many of the waterways are so rudely constructed that much waste ensues. As irrigating methods are improved this wastage will be avoided; so in assuming that a cubic foot of water will irrigate from 80 to 100 acres of land it is at the same time assumed that only the necessary amount of water will be used, and that the waterways will eventually be so constructed that the waste now almost universal will be prevented.

In determining the volume of water flowing in the streams great accuracy has not been attained. For this purpose it would be necessary to make continuous daily, or even hourly, observations for a series of years on each stream, but by the methods described in the following chapters it will be

---

[11] (continued) "cubic feet per second." Under Powell's influence, a new measurement, the acre-foot, became the standard for measuring quantity. One acre-foot is the amount of water (325,851 gallons) that will cover an acre of land to a depth of 1 foot. Establishing equivalence between old measurements of feet per second and acre-feet is at best imprecise. A diversion of 1 cubic foot per second, used continuously around the clock, would provide about 2 acre-feet per day. If system efficiency is 50% (i.e., half the water is lost in transmission to the field, which is not unusual in unlined ditch systems), then 1 acre-foot is provided for irrigation daily. Over the course of a 6-month growing season, this would amount to 180 acre-feet. If distributed over 100 acres, as Powell suggests, this amount of water would provide 1.8 feet of water per acre, enough for many crops but not enough for thirsty ones like alfalfa and cotton, which can require three times that amount. The sufficiency of Powell's recommendation becomes more questionable when one takes into account a number of other factors, including field efficiency (the amount of water turned into a field that actually becomes available for use by plants, which is always substantially less than 100%), the necessity of suspending irrigation in preparation for harvest, the difficulty of maintaining irrigation 24 hours a day, and the actuality of growing seasons shorter than 6 months in many areas of the West. Taking these matters into consideration, Powell's recommended apportionment of 1 cubic foot per second for tracts of 80 to 100 acres is Spartan indeed.

seen that a fair approximation to a correct amount has been made. For the degree of accuracy reached much is due to the fact that many of the smaller streams are already used to their fullest capacity, and thus experience has solved the problem.

Having determined from the operations of irrigation that one cubic foot per second of water will irrigate from 80 to 100 acres of land when the greatest economy is used, and having determined the volume of water or number of cubic feet per second flowing in the several streams of Utah by the most thorough methods available under the circumstances, it appears that within the territory, excluding a small portion in the southeastern corner where the survey has not yet been completed, the amount of land which it is possible to redeem by this method is about 2,262 square miles, or 1,447,920 acres.[12] Of course, this amount does not lie in a continuous body, but is scattered in small tracts along the water courses. For the purpose of exhibiting their situations a map of the territory has been prepared, and will be found accompanying this report, on which the several tracts of irrigable lands have been colored. A glance at this map will show how they are distributed. Excluding that small portion of the territory in the southeast corner not embraced in the map,[13] Utah has an area of 80,000 square miles, of which 2,262 square miles are irrigable. That is, 2.8 percent of the lands under consideration can be cultivated by utilizing all the available streams during the irrigating season.

In addition to the streams considered in this statement there are numerous small springs on the mountain sides scattered throughout the territory—springs which do not feed permanent streams; and if their waters were used for irrigation the extent of irrigable lands would be slightly increased; to what exact amount cannot be stated, but the difference would

---

[12] Powell was quite accurate. The 1997 census of the U.S. Department of Agriculture reported 1,212,201 acres under irrigation in Utah (see www.nass.dsda.gov/census/census97/volume1/volpubs.htm).

[13] Powell had not yet mapped the lands of Utah lying southeast of the Colorado River and south of the La Sal Mountains. This area includes the Abajo Mountains and all of Utah's San Juan River country.

be so small as not to materially affect the general statement, and doubtless these springs can be used in another way and to a better purpose, as will hereafter appear.

This statement of the facts relating to the irrigable lands of Utah will serve to give a clearer conception of the extent and condition of the irrigable lands throughout the Arid Region. Such as can be redeemed are scattered along the water courses, and are in general the lowest lands of the several districts to which they belong. In some of the states and territories the percentage of irrigable land is less than in Utah, in others greater, and it is probable that the percentage in the entire region is somewhat greater than in the territory which we have considered.

The Arid Region is somewhat more than four-tenths of the total area of the United States, and as the agricultural interests of so great an area are dependent upon irrigation it will be interesting to consider certain questions relating to the economy and practicability of distributing the waters over the lands to be redeemed.

Advantages of Irrigation

There are two considerations that make irrigation attractive to the agriculturist. Crops thus cultivated are not subject to the vicissitudes of rainfall; the farmer fears no droughts; his labors are seldom interrupted and his crops rarely injured by storms.[14] This immunity from drought and storm renders agricultural operations much more certain than in regions of greater humidity. Again, the water comes down from the mountains and plateaus freighted with fertilizing materials derived from the decaying vegetation and soils of the upper regions, which are spread by the flowing water over the cultivated lands. It is probable that the benefits derived from this source alone will be full compensation for the cost of the process. Hitherto these benefits have not been fully realized, from the fact that the methods

---

[14] These advantages are of course comparative. Irrigation farmers do fear the drought that reduces the amount of water available for irrigation, but they will not trade places with drought-stricken farmers who cannot irrigate at all.

employed have been more or less crude. When the flow of water over the land is too great or too rapid the fertilizing elements borne in the waters are carried past the fields, and a washing is produced which deprives the lands irrigated of their most valuable elements, and little streams cut the fields with channels injurious in diverse ways. Experience corrects these errors, and the irrigator soon learns to flood his lands gently, evenly, and economically. It may be anticipated that all the lands redeemed by irrigation in the Arid Region will be highly cultivated and abundantly productive, and agriculture will be but slightly subject to the vicissitudes of scant and excessive rainfall.

A stranger entering this Arid Region is apt to conclude that the soils are sterile, because of their chemical composition, but experience demonstrates the fact that all the soils are suitable for agricultural purposes when properly supplied with water. It is true that some of the soils are overcharged with alkaline materials, but these can in time be "washed out." Altogether the fact suggests that far too much attention has heretofore been paid to the chemical constitution of soils and too little to those physical conditions by which moisture and air are supplied to the roots of the growing plants.[15]

Coöperative Labor or Capital
Necessary for the Development of Irrigation

Small streams can be taken out and distributed by individual enterprise, but coöperative labor or aggregated capital must be employed in taking out the larger streams.

The diversion of a large stream from its channel into a system of canals demands a large outlay of labor and material. To repay this all the waters so taken out must be used, and large tracts of land thus become dependent upon a single canal. It is manifest that a farmer depending upon his own labor cannot undertake this task. To a great extent the small streams are

---

[15] With the copious application of water, alkaline soils may be leached of their salts and made productive if they are well drained.

already employed, and but a comparatively small portion of the irrigable lands can be thus redeemed; hence the chief future development of irrigation must come from the use of the larger streams. Usually the confluence of the brooks and creeks which form a large river takes place within the mountain district which furnishes its source before the stream enters the lowlands where the waters are to be used. The volume of water carried by the small streams that reach the lowlands before uniting with the great rivers, or before they are lost in the sands, is very small when compared with the volume of the streams which emerge from the mountains as rivers. This fact is important. If the streams could be used along their upper ramifications while the several branches are yet small, poor men could occupy the lands, and by their individual enterprise the agriculture of the country would be gradually extended to the limit of the capacity of the region; but when farming is dependent upon larger streams such men are barred from these enterprises until coöperative labor can be organized or capital induced to assist. Before many years all the available smaller streams throughout the entire region will be occupied in serving the lands, and then all future development will depend on the conditions above described.

In Utah Territory coöperative labor, under ecclesiastical organization, has been very successful. Outside of Utah there are but few instances where it has been tried; but at Greeley, in the State of Colorado, this system has been eminently successful.[16]

---

[16] Greeley was settled as a cooperative agricultural enterprise on 12,000 acres of railroad land in northeastern Colorado in 1870. Nathan Meeker, the agricultural editor of Horace Greeley's *New York Tribune,* organized the colony with Greeley's support. Eight years later Meeker left to become Indian agent at White River in far western Colorado, where, in 1879, his missionary zeal, lack of nerve, and intolerance of Indian ways precipitated what became known as the Meeker Massacre. It was a disaster at multiple levels. Meeker and 17 other whites at the agency were slain, and 12 white soldiers and 37 Utes perished in a battle not far away. Worse for the Utes, inevitable white demands for vengeance soon led to the permanent removal of the White River and Uncompahgre Utes from their Colorado homeland.

The Use of Smaller Streams
Sometimes Interferes with the Use of the Larger

A river emerging from a mountain region and meandering through a valley may receive small tributaries along its valley course. These small streams will usually be taken out first, and the lands which they will be made to serve will often lie low down in the valley, because the waters can be more easily controlled here and because the lands are better; and this will be done without regard to the subsequent use of the larger stream to which the smaller ones are tributary. But when the time comes to take out the larger stream, it is found that the lands which it can be made to serve lying adjacent on either hand are already in part served by the smaller streams, and as it will not pay to take out the larger stream without using all of its water, and as the people who use the smaller streams have already vested rights in these lands, a practical prohibition is placed upon the use of the larger river. In Utah, church authority, to some extent at least, adjusts these conflicting interests by causing the smaller streams to be taken out higher up in their course. Such adjustment is not so easily attained by the great body of people settling in the Rocky Mountain Region, and some provision against this difficulty is an immediate necessity. It is a difficulty just appearing, but in the future it will be one of great magnitude.

Increase of Irrigable Area by the Storage of Water

Within the Arid Region great deposits of gold, silver, iron, coal, and many other minerals are found, and the rapid development of these mining industries will demand *pari passu*[17] a rapid development of agriculture. Thus all lands that can be irrigated will be required for agricultural products necessary to supply the local market created by the mines. For this purpose the waters of the non-growing season will be stored, that they may be used in the growing season.

There are two methods of storing the waste waters. Reservoirs may be

---

[17]With equal pace or progress, or in parallel, hand in hand.

constructed near the sources of the streams and the waters held in the upper valleys, or the water may be run from the canals into ponds within or adjacent to the district where irrigation is practiced. This latter method will be employed first. It is already employed to some extent where local interests demand and favorable opportunities are afforded. In general, the opportunities for ponding water in this way are infrequent, as the depressions where ponds can easily be made are liable to be so low that the waters cannot be taken from them to the adjacent lands, but occasionally very favorable sites for such ponds may be found. This is especially true near the mountains where alluvial cones have been formed at the debouchure[18] of the streams from the mountain cañons. Just at the foot of the mountains are many places where ancient glaciation has left the general surface with many depressions favorable to ponding.

Ponding in the lower region is somewhat wasteful of water, as the evaporation is greater than above, and the pond being more or less shallow a greater proportional surface for evaporation is presented. This wastage is apparent when it is remembered that the evaporation in an arid climate may be from 60 to 80 inches annually, or even greater.[19]

Much of the waste water comes down in the spring when the streams are high and before the growing crops demand a great supply. When this water is stored the loss by evaporation will be small.

The greater storage of water must come from the construction of great reservoirs in the highland where lateral valleys may be dammed and the main streams conducted into them by canals.[20] On most streams favorable

---

[18] Where the stream exits the mouth of the canyon and enters the plain. An alluvial cone is the triangular deposit of waterborne material (mainly sand and gravel) laid down where a stream's floodwaters slow and spread. It is analogous and similar in shape to the delta of a river.

[19] In the Imperial Valley of California, probably the hottest agricultural district in the nation, annual evaporation is slightly less than 6 feet per year. Wind and humidity as well as temperature influence evaporation. At the time of Powell's writing, the empirical study of phenomena such as evaporation was only beginning; as a result, few reliable measurements were available.

[20] Powell here contemplates building reservoirs beside streams rather than athwart them. The practice was rarely followed.

sites for such water works can be found. This subject cannot be discussed at any length in a general way, from the fact that each stream presents problems peculiar to itself.

It cannot be very definitely stated to what extent irrigation can be increased by the storage of water. The rainfall is much greater in the mountain than in the valley districts. Much of this precipitation in the mountain districts falls as snow. The great snow banks are the reservoirs which hold the water for the growing seasons. Then the streams are at flood tide; many go dry after the snows have been melted by the midsummer sun; hence they supply during the irrigating time much more water than during the remainder of the year. During the fall and winter the streams are small; in late spring and early summer they are very large. A day's flow at flood time is greater than a month's flow at low water time. During the first part of the irrigating season less water is needed, but during that same time the supply is greatest. The chief increase will come from the storage of this excess of water in the early part of the irrigating season. The amount to be stored will then be great, and the time of this storage will be so short that it will be but little diminished by evaporation. The waters of the fall and winter are so small in amount that they will not furnish a great supply, and the time for their storage will be so great that much will be lost by evaporation. The increase by storage will eventually be important, and it would be wise to anticipate the time when it will be needed by reserving sites for principal reservoirs and larger ponds.[21]

Timber Lands

Throughout the Arid Region timber of value is found growing spontaneously on the higher plateaus and mountains. These timber regions are

---

[21]Another future point of contention. Powell recommended that such sites be withdrawn from eligibility for entry under the Homestead Act and other public land laws. Because such lands were often attractive for settlement in their own right or desirable for the purpose of achieving strategic control over a watershed (a rancher, for instance, might file on such an area to forestall homesteading downstream), Powell's recommendation met strong opposition from many sectors.

bounded above and below by lines which are very irregular, due to local conditions. Above the upper line no timber grows because of the rigor of the climate, and below no timber grows because of aridity. Both the upper and lower lines descend in passing from south to north; that is, the timber districts are found at a lower altitude in the northern portion of the Arid Region than in the southern. The forests are chiefly of pine, spruce, and fir, but the pines are of principal value. Below these timber regions, on the lower slopes of mountains, on the mesas and hills, low, scattered forests are often found, composed mainly of dwarfed piñon pines and cedars.[22] These stunted forests have some slight value for fuel, and even for fencing, but the forests of principal value are found in the Timber Region as above described.

Primarily the growth of timber depends on climatic conditions—humidity and temperature. Where the temperature is higher, humidity must be greater, and where the temperature is lower, humidity may be less. These two conditions restrict the forests to the highlands, as above stated. Of the two factors involved in the growth of timber, that of the degree of humidity is of the first importance; the degree of temperature affects the problem comparatively little, and for most of the purposes of this discussion may be neglected. For convenience, all these upper regions where conditions of temperature and humidity are favorable to the growth of timber may be called the *timber regions.*

Not all these highlands are alike covered with forests. The timber regions are only in part *areas of standing timber.* This limitation is caused by fire. Throughout the timber regions of all the arid land fires annually destroy larger or smaller districts of timber, now here, now there, and this destruction is on a scale so vast that the amount taken from the lands for industrial purposes sinks by comparison into insignificance. The cause of this great destruction is worthy of careful attention. The conditions under which these fires rage are climatic. Where the rainfall is great and extreme droughts are infrequent, forests grow with-

---

[22] *Pinus edulis* and various juniper species.

out much interruption from fires; but between that degree of humidity necessary for their protection, and that smaller degree necessary to growth, all lands are swept bare by fire to an extent which steadily increases from the more humid to the more arid districts, until at last all forests are destroyed, though the humidity is still sufficient for their growth if immunity from fire were secured. The amount of mean annual rainfall necessary to the growth of forests if protected from fire is probably about the same as the amount necessary for agriculture without irrigation; at any rate, it is somewhere from 20 to 24 inches. All timber growth below that amount is of a character so stunted as to be of little value, and the growth is so slow that, when once the timber has been taken from the country, the time necessary for a new forest growth is so great that no practical purpose is subserved.[23]

The evidence that the growth of timber, if protected from fires, might be extended to the limits here given is abundant. It is a matter of experience that planted forests thus protected will thrive throughout the prairie region and far westward on the Great Plains. In the mountain region it may be frequently observed that forest trees grow low down on the mountain slopes and in the higher valleys wherever local circumstances protect them from fires, as in the case of rocky lands that give insufficient footing to the grass and shrubs in which fires generally spread.[24] These cases must not be confounded with those patches of forest that grow on alluvial cones where rivers leave mountain cañons and enter valleys or plains. Here the streams, clogged by the material washed from the adjacent mountains by storms, are frequently turned from their courses and divided into many channels run-

---

[23] Powell was exactly on the mark: "The 500-mm (20-inch) precipitation line roughly delimits the boundary east of which it is possible to grow trees without irrigation or access to ground water" (Spurr and Barnes, *Forest Ecology*, 272–73). The 20-inch isohyet also roughly delimits the commercially valuable ponderosa pine zone from the piñon–juniper woodland (Powell's "stunted forests"). For more on Powell's views on forests and fire, see Selection 13, "The Non-Irrigable Lands of the Arid Region," and accompanying notes.

[24] Powell here recognizes the importance of "fine fuels"—grasses and other herbaceous material—for carrying fire into or through a forest stand. The natural demarcation between prairie and woodland vegetation is commonly determined by the dynamic Powell is describing.

ning near the surface. Thus a subterranean watering is effected favorable to the growth of trees, as their roots penetrate to sufficient depth. Usually this watering is too deep for agriculture, so that forests grow on lands that cannot be cultivated without irrigation.

Fire is the immediate cause of the lack of timber on the prairies, the eastern portion of the Great Plains, and on some portions of the highlands of the Arid Region; but fires obtain their destructive force through climatic conditions, so that directly and remotely climate determines the growth of all forests. Within the region where prairies, groves, and forests appear, the local distribution of timber growth is chiefly dependent upon drainage and soil, a subject which needs not be here discussed. Only a small portion of the Rocky Mountain Region is protected by climatic conditions from the invasion of fires, and a sufficiency of forests for the country depends upon the control which can be obtained over that destructive agent. A glance at the map of Utah will exhibit the extent and distribution of the forest region throughout that territory, and also show what portions of it are in fact occupied by standing timber. The *area of standing timber,* as exhibited on the map, is but a part of the Timber Region as there shown, and includes all of the timber, whether dense or scattered.

Necessarily the area of standing timber has been generalized. It was not found practicable to indicate the growth of timber in any refined way by grading it, and by rejecting from the general area the innumerable small open spaces. If the area of standing timber were considered by acres, and all acres not having timber valuable for milling purposes rejected, the extent would be reduced at least to one-fourth of that colored. Within the territory represented on the map the Timber Region has an extent of 18,500 square miles; that is, 23 percent belongs to the Timber Region.[25] The general area of standing timber is about 10,000 square miles, or 12.5 percent of the entire area. The area of milling timber, determined in the more refined way indi-

---

[25] The most surprising feature of this map, which is reproduced in this volume as maps 3 and 4, is the depiction of the "Timber Region." In the original color-shaded edition of the map, the area of "Standing Timber" is shown in dark green, corresponding to the 10,000

cated above, is about 2,500 square miles, or $3^{1}/8$ percent of the area embraced on the map. In many portions of the Arid Region these percentages are much smaller. This is true of southern California, Nevada, southern Arizona, and Idaho. In other regions the percentages are larger. Utah gives about a fair average. In general it may be stated that the timber regions are fully adequate to the growth of all the forests which the industrial interests of the country will require if they can be protected from desolation by fire. No limitation to the use of the forests need be made. The amount which the citizens of the country will require will bear but a small proportion to the amount which the fires will destroy; and if the fires are prevented, the renewal by annual growth will more than replace that taken by man. The protection of the forests of the entire Arid Region of the United States is reduced to one single problem—Can these forests be saved from fire? The writer has witnessed two fires in Colorado,[26] each of which destroyed more timber than all that used by the citizens of that State from its settlement to the present day; and at least three in Utah, each of which has destroyed more timber than that taken by the people of the territory since its occupation. Similar fires have been witnessed by other members of the surveying corps. Everywhere throughout the Rocky Mountain Region the explorer away from the beaten paths of civilization meets with great areas of dead forests; pines with naked arms and charred trunks attesting to the former presence of

---

[25] (*continued*) square miles referenced here. An area of nearly equal size (amounting to 8,500 square miles, by Powell's math) is shown in tan and described by the legend as "Area destitute of Timber on account of Fires." This area appears to include a great deal of land lower in elevation than the standing timber lands, which poses something of a puzzle: One would expect a majority of lower-elevation timber land to be ponderosa pine savanna or piñon–juniper woodland, both of which systems would have been well adapted to fire in 1878 by reason of experiencing frequent low-intensity burns in their pregrazing, pre-fire–suppression condition. Considering their adaptation, it seems unlikely that any observer as discerning as Powell would describe them as "destitute of timber on account of fires." In any event, the vastness of the area so depicted on the map seems greater than even the most pyromaniacal historical reconstruction would show.

[26] Powell in fact started one of them. See his description in Selection 13, "The Non-Irrigable Lands of the Arid Region."

this greater destroyer. The younger forests are everywhere beset with falling timber, attesting to the rigor of the flames, and in seasons of great drought the mountaineer sees the heavens filled with clouds of smoke.

In the main these fires are set by Indians. Driven from the lowlands by advancing civilization, they resort to the higher regions until they are forced back by the deep snows of winter. Want, caused by the restricted area to which they resort for food; the desire for luxuries to which they were strangers in their primitive condition, and especially the desire for personal adornment, together with a supply of more effective instruments for hunting and trapping, have in late years, during the rapid settlement of the country since the discovery of gold and the building of railroads, greatly stimulated the pursuit of animals for their furs—the wealth and currency of the savage. On their hunting excursions they systematically set fire to forests for the purpose of driving the game. This is a fact well known to all mountaineers. Only the white hunters of the region properly understand why these fires are set, it being usually attributed to a wanton desire on the part of the Indians to destroy that which is of value to the white man. The fires can, then, be very greatly curtailed by the removal of the Indians.[27]

These forest regions are made such by inexorable climatic conditions. They are high among the summer frosts. The plateaus are scored by deep cañons, and the mountains are broken with crags and peaks. Perhaps at some distant day a hardy people will occupy little glens and mountain valleys, and wrest from an unwilling soil a scanty subsistence among the rigors of a sub-arctic climate. Herdsmen having homes below may in the summer time drive their flocks to the higher lands to crop the scanty herbage.

---

[27] Powell's ideas on the causation of fire later changed, and he eventually attributed most fires to the activities of white settlers. Although here he tends to overlook lightning as a source of ignition and seems unaware that many forest types (ponderosa and lodgepole pine, for instance) are well adapted to fire, the point he raises is an important yet unsettled one in the environmental history of the West. Did an epidemic of stand-changing fire take place in the mountains of the West in the second half of the nineteenth century, as Powell's observations suggest? And if it did, did it arise from natural (possibly climatic) causes or as a result of human activities associated with colonization and settlement, as Powell also suggests? Or did it arise from the convergence of these factors and possibly others?

Where mines are found mills will be erected and little towns spring up, but in general habitations will be remote. The forests will be dense here or scattered there, as the trees may with ease or difficulty gain a foothold, but the forest regions will remain such, to be stripped of timber here and there from time to time to supply the wants of the people who live below; but once protected from fires, the forests will increase in extent and value. The first step to be taken for their protection must be by prohibiting the Indians from resorting thereto for hunting purposes, and then slowly, as the lower country is settled, the grasses and herbage of the highlands, in which fires generally spread, will be kept down by summer pasturage,[28] and the dead and fallen timber will be removed to supply the wants of people below. This protection, though sure to come at last, will be tardy, for it depends upon the gradual settlement of the country; and this again depends upon the development of the agricultural and mineral resources and the establishment of manufactories, and to a very important extent on the building of railroads, for the whole region is so arid that its streams are small, and so elevated above the level of the sea that its few large streams descend too rapidly for navigation.

Agricultural and Timber Industries Differentiated

It is apparent that the irrigable lands are more or less remote from the timber lands; and as the larger streams are employed for irrigation, in the future the extended settlements will be still farther away. The pasturage lands that in a general way intervene between the irrigable and timber lands have a scanty supply of dwarfed forests, as already described, and the people in occupying these lands will not resort, to any great extent, to the mountains for timber; hence timber and agricultural enterprises will be more or less differentiated; lumbermen and woodmen will furnish to the people below their supply of building and fencing material and fuel. In some cases it will

---

[28] Again, Powell's prediction was accurate, if the result was less desirable than he appreciated. The incidence of periodic, low-intensity wildfire, as reconstructed from the growth rings of fire-scarred trees, consistently ceases area by area with the initiation of widespread domestic livestock grazing.

be practicable for the farmers to own their timber lands, but in general the timber will be too remote, and from necessity such a division of labor will ensue.

Cultivation of Timber

In the irrigable districts much timber will be cultivated along the canals and minor waterways. It is probable that in time a sufficient amount will thus be raised to supply the people of the irrigable districts with fuel wherever such fuel is needed, but often such a want will not exist, for in the Rocky Mountain Region there is a great abundance of lignitic coals that may be cheaply mined. All these coals are valuable for domestic purposes, and many superior grades are found. These coals are not uniformly distributed, but generally this source of fuel is ample.

Pasturage Lands

The irrigable lands and timber lands constitute but a small fraction of the Arid Region. Between the lowlands on the one hand and the highlands on the other is found a great body of valley, mesa, hill, and low mountain lands. To what extent, and under what conditions can they be utilized? Usually they bear a scanty growth of grasses. These grasses are nutritious and valuable both for summer and winter pasturage. Their value depends upon peculiar climatic conditions; the grasses grow to a great extent in scattered bunches, and mature seeds in larger proportion perhaps than the grasses of the more humid regions. In general the winter aridity is so great that the grasses when touched by the frosts are not washed down by the rains and snows to decay on the moist soil, but stand firmly on the ground all winter long and "cure," forming a *quasi* uncut hay. Thus the grass lands are of value both in summer and winter. In a broad way, the greater or lesser abundance of the grasses is dependent on latitude and altitude; the higher the latitude the better are the grasses, and they improve as the altitude increases. In very low altitudes and latitudes the grasses are so scant as to be of no value; here the true deserts are found. These conditions obtain in southern California, southern Nevada, southern Arizona, and southern

New Mexico, where broad reaches of land are naked of vegetation, but in ascending to the higher lands the grass steadily improves. Northward the deserts soon disappear, and the grass becomes more and more luxuriant to our northern boundary. In addition to the desert lands mentioned, other large deductions must be made from the area of the pasturage lands. There are many districts in which the "country rock" is composed of incoherent sands and clays; sometimes sediments of ancient Tertiary lakes; elsewhere sediments of more ancient Cretaceous seas. In these districts perennial or intermittent streams have carved deep waterways, and the steep hills are ever washed naked by fierce but infrequent storms, as the incoherent rocks are unable to withstand the beating of the rain. These districts are known as the *mauvaises terres* or bad lands of the Rocky Mountain Region. In other areas the streams have carved labyrinths of deep gorges and the waters flow at great depths below the general surface. The lands between the streams are beset with towering cliffs, and the landscape is an expanse of naked rock. These are the alcove lands and cañon lands of the Rocky Mountain Region. Still other districts have been the theater of late volcanic activity, and broad sheets of naked lava are found; cinder cones are frequent, and scoria[29] and ashes are scattered over the land. These are the lava-beds of the Rocky Mountain Region. In yet other districts, low broken mountains are found with rugged spurs and craggy crests. Grasses and chaparral grow among the rocks, but such mountains are of little value for pasturage purposes.

After making all the deductions, there yet remain vast areas of valuable pasturage land bearing nutritious but scanty grass. The lands along the creeks and rivers have been relegated to that class which has been described as irrigable, hence the lands under consideration are away from the permanent streams. No rivers sweep over them and no creeks meander among their hills.

Though living water is not abundant, the country is partially supplied by scattered springs, that often feed little brooks whose waters never join

---

[29] A cinderlike, cellular lava.

the great rivers on their way to the sea, being able to run but a short distance from their fountains, when they spread among the sands to be reëvaporated. These isolated springs and brooks will in many cases furnish the water necessary for the herds that feed on the grasses. When springs are not found wells may be sometimes dug, and where both springs and wells fail reservoirs may be constructed.[30] Wherever grass grows water may be found or saved from the rains in sufficient quantities for all the herds that can live on the pasturage.

Pasturage Farms Need Small Tracts of Irrigable Land

The men engaged in stock raising need small areas of irrigable lands for gardens and fields where agricultural products can be raised for their own consumption, and where a store of grain and hay may be raised for their herds when pressed by the severe storms by which the country is sometimes visited. In many places the lone springs and streams are sufficient for these purposes. Another and larger source of water for the fertilization of the gardens and fields of the pasturage farms is found in the smaller branches and upper ramifications of the larger irrigating streams. These brooks can be used to better advantage for the pasturage farms as a supply of water for stock gardens and small fields than for farms where agriculture by irrigation is the only industry. The springs and brooks of the permanent drainage can be employed in making farms attractive and profitable where large herds may be raised in many great districts throughout the Rocky Mountain Region.

The conditions under which these pasturage lands can be employed are worthy of consideration.

The Farm Unit for Pasturage Lands

The grass is so scanty that the herdsman must have a large area for the support of his stock. In general a quarter section of land alone is of no value to

---

[30] Stock tanks, in which the West abounds.

him; the pasturage it affords is entirely inadequate to the wants of a herd that the poorest man needs for his support.[31]

Four square miles may be considered as the minimum amount necessary for a pasturage farm, and a still greater amount is necessary for the larger part of the lands; that is, pasturage farms, to be of any practicable value, must be of at least 2,560 acres, and in many districts they must be much larger.[32]

Regular Division Lines for Pasturage Farms Not Practicable

Many a brook which runs but a short distance will afford sufficient water for a number of pasturage farms; but if the lands are surveyed in regular tracts as square miles or townships, all the water sufficient for a number of pasturage farms may fall entirely within one division.[33] If the lands are thus surveyed, only the divisions having water will be taken, and the farmer obtaining title to such a division or farm could practically occupy all the country adjacent by owning the water necessary to its use. For this reason divisional

---

[31]Powell is referring to the 160-acre homestead, which the Preemption Act of 1841 and the Homestead Act of 1862 enshrined as a national standard.

[32] Powell adds this footnote: "For the determination of the proper unit for pasturage farms the writer has conferred with many persons living in the Rocky Mountain Region who have had experience. His own observations have been extensive, and for many years while conducting surveys and making long journeys through the Arid Region this question has been uppermost in his mind. He fears that this estimate will disappoint many of his western friends, who will think he has placed the minimum too low, but after making the most thorough examination of the subject possible he believes the amount to be sufficient for the best pasturage lands, especially such as are adjacent to the minor streams of the general drainage, and when these have been taken by actual settlers the size of the pasturage farms may be increased as experience proves necessary."

[33] Powell is objecting to the rectilinear survey of aliquot parts, by which lands of a state or territory were divided into a vast checkerboard of townships 6 miles on a side and containing 36 sections, each a square mile (640 acres). Sections might be further subdivided into quarter sections (160 acres, e.g., the northwest quarter of a section), quarter-quarters (40 acres, the southwest quarter of the northwest quarter of a section), or still smaller square and rectangular units. The virtue of the system was the speed with which it could be established within a region and the ease and simplicity with which lands might then be described. Its liability, as Powell points out, was that the boundaries it created were geometric abstractions that bore no relationship to the character of the land being partitioned.

surveys should conform to the topography, and be so made as to give the greatest number of water fronts. For example, a brook carrying water sufficient for the irrigation of 200 acres of land might be made to serve for the irrigation of 20 acres to each of ten farms, and also supply the water for all the stock that could live on ten pasturage farms, and ten small farmers could have homes. But if the water was owned by one man, nine would be excluded from its benefits and nine-tenths of the land remain in the hands of the government.[34]

Farm Residences Should Be Grouped

These lands will maintain but a scanty population. The homes must necessarily be widely scattered from the fact that the farm unit must be large. That the inhabitants of these districts may have the benefits of the local social organizations of civilization—as schools, churches, etc., and the benefits of coöperation in the construction of roads, bridges, and other local improvements, it is essential that the residences should be grouped to the greatest possible extent.[35] This may be practically accomplished by making the pasturage farms conform to topographic features in such manner as to give the greatest possible number of water fronts.

Pasturage Lands Cannot Be Fenced

The great areas over which stock must roam to obtain subsistence usually prevents the practicability of fencing the lands. It will not pay to fence the pasturage fields, hence in many cases the lands must be occupied by herds roaming in common; for poor men coöperative pasturage is necessary, or

---

[34] As Powell predicted, many western ranches were built by controlling water sources and thereby controlling use of adjacent public domain.

[35] Isolation was one of the greatest burdens of ranch life, and of agricultural life in horse-drawn America generally. Powell is here suggesting at least a partial remedy. The proponents of irrigation agriculture carried this argument still further, emphasizing that in desert areas with long growing seasons, irrigation made possible productive farms of very small size, which in turn would enable families to live close together and to cultivate manners as well as the soil.

communal regulations for the occupancy of the ground and for the division of the increase of the herds. Such communal regulations have already been devised in many parts of the country.[36]

RECAPITULATION

The Arid Region of the United States is more than four-tenths of the area of the entire country excluding Alaska.

In the Arid Region there are three classes of lands, namely, irrigable lands, timber lands, and pasturage lands.

Irrigable Lands

Within the Arid Region agriculture is dependent upon irrigation.

The amount of irrigable land is but a small percentage of the whole area.

The chief development of irrigation depends upon the use of the large streams.

For the use of large streams coöperative labor or capital is necessary.

The small streams should not be made to serve lands so as to interfere with the use of the large streams.

Sites for reservoirs should be set apart, in order that no hindrance may be placed upon the increase of irrigation by the storage of water.

Timber Lands

The timber regions are on the elevated plateaus and mountains.

The timber regions constitute from 20 to 25 percent of the Arid Region.

---

[36] Powell may have been thinking of the Hispanic villages of northern New Mexico, among other places. These villages were generally established within Spanish or Mexican land grants, which had been awarded to groups of settlers by sovereign authority. They took form as clustered settlements near irrigated lands. Within them, individual families owned their house sites and irrigated fields privately, but the grazing and forest lands that comprised the rest of the land grant functioned as a commons in which all residents of the village had rights of use for pasturing livestock, wood cutting, and other subsistence needs. The communal regulations governing use of the commons were rarely if ever codified. They consisted instead of traditions and accepted patterns of use.

The area of standing timber is much less than the timber region, as the forests have been partially destroyed by fire.

The timber regions cannot be used as farming lands; they are valuable for forests only.

To preserve the forests they must be protected from fire. This will be largely accomplished by removing the Indians.

The amount of timber used for economic purposes will be more than replaced by the natural growth.

In general the timber is too far from the agricultural lands to be owned and utilized directly by those who carry on farming by irrigation.

A division of labor is necessary, and special timber industries will be developed, and hence the timber lands must be controlled by lumbermen and woodmen.

Pasturage Lands

The grasses of the pasturage land are scant, and the lands are of value only in large quantities.

The farm unit should not be less than 2,560 acres.

Pasturage farms need small tracts of irrigable land; hence the small streams of the general drainage system and the lone springs and streams should be reserved for such pasturage farms.

The division of these lands should be controlled by topographic features in such manner as to give the greatest number of water fronts to the pasturage farms.

Residences of the pasturage farms should be grouped, in order to secure the benefits of local social organizations, and coöperation in public improvements.

The pasturage lands will not usually be fenced, and hence herds must roam in common.

As the pasturage lands should have water fronts and irrigable tracts, and as the residences should be grouped, and as the lands cannot be economically fenced and must be kept in common, local communal regulations or coöperation is necessary.

The Land System Needed for the Arid Region

The growth and prosperity of the Arid Region will depend largely upon a land system which will comply with the requirements of the conditions and facts briefly set forth in the former chapter.

Any citizen of the United States may acquire title to public lands by purchase at public sale or by ordinary "private entry," and in virtue of preëmption, homestead, timber culture, and desert land laws.[1]

Purchase at public sale may be effected when the lands are offered at public auction to the highest bidder, either pursuant to proclamation by the President or public notice given in accordance with instructions from the General Land Office. If the land is thus offered and purchasers are not found, they are then subject to "private entry" at the rate of $1.25 or $2.50 per acre. For a number of years it has not been the practice of the Government to dispose of the public lands by these methods; but the public lands of the southern states are now, or soon will be, thus offered for sale.

Any citizen may preëmpt 160 acres of land, and by settling thereon, erecting a dwelling, and making other improvements, and by paying $1.25 per acre in some districts, without the boundaries of railroad grants, and $2.50 within the boundaries of railroad grants in others, may acquire title thereto. The preëmption right can be exercised but once. No person can

---

[1] In this selection, as a prelude to recommending changes in the nation's land laws, Powell reviews the formidable array of processes by which land might be removed from the public domain and conveyed to private ownership. He reviews the provisions of various preemption acts, the Homestead Act (1862), the Timber Culture Act (1873), the Desert Land Act (1877), and other measures. For a review of public land law in the nineteenth century, see Hibbard, *A History of the Public Land Policies.*

exercise the preëmption right who is already the owner of 320 acres of land.

Any citizen may, under the homestead privilege, obtain title to 160 acres of land valued at $1.25 per acre, or 80 acres valued at the rate of $2.50, by payment of $5 in the first case and $10 in the last, and by residing on the land for the term of five years and by making certain improvements.

The time of residence is shortened for persons who have served in the army or navy of the United States, and any such person may homestead 160 acres of land valued at $2.50 per acre.

Any citizen may take advantage of both the homestead and preëmption privileges.

Under the timber culture act, any citizen who is the head of a family may acquire title to 160 acres of land in the prairie region by cultivating timber thereon in certain specific quantities; the title can be acquired at the expiration of eight years from the date of entry.

Any citizen may acquire title to one section of desert land (irrigable lands as described in this paper) by the payment at the time of entry of 25 cents per acre, and by redeeming the same by irrigation within a period of three years and by the payment of $1 per acre at the expiration of that time, and a patent will then issue.

Provision is also made for the disposal of public lands as town sites.

From time to time land warrants have been issued by the Government as bounties to soldiers and sailors, and for other purposes. These land warrants have found their way into the market, and the owners thereof are entitled to enter Government lands in the quantities specified in the warrants.

Agricultural scrip has been issued for the purpose of establishing and endowing agricultural schools. A part of this scrip has been used by the schools in locating lands for investment. Much of the scrip has found its way into the market and is used by private individuals. Warrants and scrip can be used when lands have been offered for sale, and preëmptors can use them in lieu of money.

Grants of lands have been made to railroad and other companies, and as these railroads have been completed in whole or in part, the companies have obtained titles to the whole or proportional parts of the lands thus granted.

Where the railroads are unfinished the titles are inchoate to an extent proportional to the incomplete parts.

With small exceptions, the lands of the Arid Region have not been offered for sale at auction or by private entry.

The methods, then, by which the lands under consideration can be obtained from the Government are by taking advantage of the preëmption, homestead, timber culture, or desert land privileges.

Irrigable Lands

By these methods adequate provision is made for actual settlers on all irrigable lands that are dependent on the waters of minor streams; but these methods are insufficient for the settlement of the irrigable lands that depend on the larger streams, and also for the pasturage lands and timber lands, and in this are included nearly all the lands of the Arid Region. If the irrigable lands are to be sold, it should be in quantities to suit purchasers, and but one condition should be imposed, namely, that the lands should be actually irrigated before the title is transferred to the purchaser. This method would provide for the redemption of these lands by irrigation through the employment of capital. If these lands are to be reserved for actual settlers, in small quantities, to provide homes for poor men, on the principle involved in the homestead laws, a general law should be enacted under which a number of persons would be able to organize and settle on irrigable districts, and establish their own rules and regulations for the use of the water and subdivision of the lands, but in obedience to the general provisions of the law.[2]

Timber Lands

The timber lands cannot be acquired by any of the methods provided in the preëmption, homestead, timber culture, and desert land laws, from the fact

---

[2] The theme announced here, integral to Powell's specific proposals, is to develop land laws hospitable to settlement by both groups and individuals.

that they are not agricultural lands. Climatic conditions make these methods inoperative. Under these laws "dummy entries" are sometimes made. A man wishing to obtain the timber from a tract of land will make homestead or preëmption entries by himself or through his employees without intending to complete the titles, being able thus to hold these lands for a time sufficient to strip them of their timber.

This is thought to be excusable by the people of the country, as timber is necessary for their industries, and the timber lands cannot honestly be acquired by those who wish to engage in timber enterprises. Provision should be made by which the timber can be purchased by persons or companies desiring to engage in the lumber or wood business, and in such quantities as may be necessary to encourage the construction of mills, the erection of flumes, the making of roads, and other improvements necessary to the utilization of the timber for the industries of the country.

Pasturage Lands

If divisional surveys were extended over the pasturage lands, favorable sites at springs and along small streams would be rapidly taken under the homestead and preëmption privileges for the nuclei of pasturage farms.

Unentered lands contiguous to such pasturage farms could be controlled to a greater or less extent by those holding the water, and in this manner the pasturage of the country would be rendered practicable. But the great body of land would remain in the possession of the Government; the farmers owning the favorable spots could not obtain possession of the adjacent lands by homestead or preëmption methods, and if such adjacent lands were offered for sale, they could not afford to pay the Government price.

Certain important facts relating to the pasturage farms may be advantageously restated.

The farm unit should not be less than 2,560 acres; the pasturage farms need small bodies of irrigable land; the division of these lands should be controlled by topographic features to give water fronts; residences of the pasturage lands should be grouped; the pasturage farms cannot be fenced—they must be occupied in common.

The homestead and preëmption methods are inadequate to meet these conditions. A general law should be enacted to provide for the organization of pasturage districts, in which the residents should have the right to make their own regulations for the division of the lands, the use of the water for irrigation and for watering the stock, and for the pasturage of the lands in common or in severalty.[3] But each division or pasturage farm of the district should be owned by an individual; that is, these lands could be settled and improved by the "colony" plan better than by any other. It should not be understood that the colony system applies only to such persons as migrate from the east in a body; any number of persons already in this region could thus organize. In fact very large bodies of these lands would be taken by people who are already in the country and who have herds with which they roam about seeking water and grass, and making no permanent residences and no valuable improvements. Such a plan would give immediate relief to all these people.

This district or colony system is not untried in this country. It is essentially the basis of all the mining district organizations of the west. Under it the local rules and regulations for the division of mining lands, the use of water, timber, etc., are managed better than they could possibly be under specific statutes of the United States. The association of a number of peo-

---

[3] More than a quarter century passed before the first pasturage districts in the form of grazing allotments were established on lands of the young system of National Forests that was created after Powell's death. Another quarter century passed before Congress finally created a system of grazing leases for the remaining public domain. In the meantime, these lands suffered intensely from overgrazing in an immense tragedy of the commons. At the heart of the tragedy was the inability of ranchers to retain the use of land unless they stocked it fully: Any unstocked land might be occupied by the herd of another rancher. As a result, rangelands were rarely rested and experienced continuous overuse. Powell's proposal would have empowered settlers to achieve ownership of rangeland, to defend it against competitors, and to regulate its use. This would by no means have eliminated the possibility of overuse, but it would have laid the foundation for ranchers, guided by self-interest, to conserve the resource on which they depended.

ple prevents single individuals from having undue control of natural privileges, and secures an equitable division of mineral lands; and all this is secured in obedience to statutes of the United States providing general regulations.

Customs are forming and regulations are being made by common consent among the people in some districts already; but these provide no means for the acquirement of titles to land, no incentive is given to the improvement of the country, and no legal security to pasturage rights.

If, then, the irrigable lands can be taken in quantities to suit purchasers, and the colony system provided for poor men who wish to coöperate in this industry; if the timber lands are opened to timber enterprises, and the pasturage lands offered to settlement under a colony plan like that indicated above, a land system would be provided for the Arid Region adapted to the wants of all persons desiring to become actual settlers therein. Thousands of men who now own herds and live a semi-nomadic life; thousands of persons who now roam from mountain range to mountain range prospecting for gold, silver, and other minerals; thousands of men who repair to that country and return disappointed from the fact that they are practically debarred from the public lands; and thousands of persons in the eastern states without employment, or discontented with the rewards of labor, would speedily find homes in the great Rocky Mountain Region.

In making these recommendations, the wisdom and beneficence of the homestead system have been recognized and the principles involved have been considered paramount.

To give more definite form to some of the recommendations for legislation made above, two bills have been drawn, one relating to the organization of irrigation districts, the other to pasturage districts. These bills are presented here. It is not supposed that these forms are the best that could be adopted; perhaps they could be greatly improved; but they have been carefully considered, and it is believed they embody the recommendations made above.

A Bill to Authorize the Organization of Irrigation Districts by Homestead Settlements upon the Public Lands Requiring Irrigation for Agricultural Purposes

Be it enacted by the Senate and House of Representatives of the United States of America in Congress assembled, That it shall be lawful for any nine or more persons who may be entitled to acquire a homestead from the public lands, as provided for in sections twenty-two hundred and eighty-nine to twenty-three hundred and seventeen, inclusive, of the Revised Statutes of the United States, to settle an irrigation district and to acquire titles to irrigable lands under the limitations and conditions hereinafter provided.

Sec. 2. That it shall be lawful for the persons mentioned in section one of this act to organize an irrigation district in accordance with a form and general regulations to be prescribed by the Commissioner of the General Land Office, which shall provide for a recorder; and said persons may make such by-laws, not in conflict with said regulations, as they may deem wise for the use of waters in such district for irrigation or other purposes, and for the division of the lands into such parcels as they may deem most convenient for irrigating purposes; but the same must accord with the provisions of this act.

Sec. 3. That all lands in those portions of the United States where irrigation is necessary to agriculture, which can be redeemed by irrigation and for which there is accessible water for such purpose, not otherwise utilized or lawfully claimed, sufficient for the irrigation of three hundred and twenty acres of land, shall, for the purposes set forth in this act, be classed as irrigable lands.[4]

---

[4] The task of classification would be immense. Without approving anything similar to this bill, Congress finally authorized Powell to undertake the classification of irrigable lands in 1888, but his efforts never escaped the storm of controversy that immediately attached to them. Western interests perceived that Powell would arrest all western settlement for an indefinite time while his painstaking and meticulous survey proceeded. In 1890, the budget of the Irrigation Survey was slashed to $165,000, making completion of its task impossible (Goetzmann, *Exploration and Empire,* 596–99).

Sec 4. That it shall be lawful for the requisite number of persons, as designated in section one of this act, to select from the public lands designated as irrigable lands in section three of this act, for the purpose of settling thereon, an amount of land not exceeding eighty acres to each person; but the lands thus selected by the persons desiring to organize an irrigation district shall be in one continuous tract, and the same shall be subdivided as the regulations and by-laws of the irrigation district shall prescribe: *Provided,* That no one person shall be entitled to more than eighty acres.[5]

Sec. 5. That whenever such irrigation district shall be organized the recorder of such district shall notify the register and receiver of the land district in which such irrigation district is situate, and also the Surveyor-General of the United States,[6] that such irrigation district has been organized; and each member of the organization of said district shall file a declaration with the register and receiver of said land district that he has settled upon a tract of land within such irrigation district, not exceeding the prescribed amount, with the intention of residing thereon and obtaining a title thereto under the provisions of this act.

Sec. 6. That if within three years after the organization of the irrigation district the claimants therein, in their organized capacity, shall apply for a survey of said district to the Surveyor-General of the United States, he shall cause a proper survey to be made, together with a plat of the same; and on

---

[5] The Newlands Act of 1902, which established the Reclamation Service (later the Bureau of Reclamation), was the first federal legislation to incorporate elements of Powell's plan. It provided for irrigated farms of not more than 160 acres, a limit more honored in the breach than the observance. The Newlands Act also authorized the government to finance and build dams, a major departure from Powell's recommendations. Powell believed that dam construction should be the work of local communities and private capital. The entry of the federal government into this arena set the stage for projects, such as Hoover Dam, that were far more grandiose than those Powell contemplated and that embodied a concentration of economic and political power inherently hostile to his vision of community-based democracy.

[6] A surveyor-general was appointed for each western territory and was responsible for executing its rectilinear survey. The surveyors-general of the territories reported to the surveyor-general of the United States, who was an officer of the General Land Office in the Department of the Interior. The General Land Office was headed by a commissioner, referred to in Sec. 6.

this plat each tract or parcel of land into which the district is divided, such tract or parcel being the entire claim of one person, shall be numbered, and the measure of every angle, the length of every line in the boundaries thereof, and the number of acres in each tract or parcel shall be inscribed thereon, and the name of the district shall appear on the plat in full; and this plat and the field-notes of such survey shall be submitted to the Surveyor-General of the United States; and it shall be the duty of that officer to examine the plat and notes therewith and prove the accuracy of the survey in such manner as the Commissioner of the General Land Office may prescribe; and if it shall appear after such examination and proving that correct surveys have been made, and that the several tracts claimed are within the provisions of this act, he shall certify the same to the register of the land district, and shall thereupon furnish to the said register of the land district, and to the recorder of the irrigation district, and to the recorder or clerk of the county in which the irrigation district is situate, and to the Commissioner of the General Land Office, a copy thereof to each, and the original shall be retained in the office of the Surveyor-General of the United States for preservation.

Sec. 7. That each person applying for the benefits of this act shall, in addition to compliance therewith, conform to the methods provided for the acquirement of a homestead in sections twenty-two hundred and eighty-nine to twenty-three hundred and seventeen, inclusive, of the Revised Statutes of the United States, so far as they are applicable and consistent with this act, and shall also furnish such evidence as the commissioner of the General Land Office may require that such land has actually been redeemed by irrigation, and may thereupon obtain a patent:[7] *Provided,* That no person shall obtain a patent under this act to any coal lands, town sites, or tracts of public lands on which towns may

---

[7] A patent is a title. Untitled land remains in the domain of the sovereign, which in the United States is the federal government. Until the government issues a patent to land (the physical instrument of which commonly is framed in purple, the color of the sovereign), private title does not exist and cannot be transferred between seller and buyer. In theory a chain of title must originate with a patent or its equivalent.

have been built, or to any mine of gold, silver, cinnabar,[8] copper, or other mineral for the sale or disposal of which provision has been made by law.

Sec. 8. That the lands patented under the provisions of this act shall be described as irrigation farms, and designated by the number of the tract or parcel and the name of the irrigation district.

Sec. 9. That the right to the water necessary to the redemption of an irrigation farm shall inhere in the land from the time of the organization of the irrigation district, and in all subsequent conveyances the right to the water shall pass with the title to the land.[9] But if after the lapse of five years from the date of the organization of the district the owner of any irrigation farm shall have failed to irrigate the whole or any part of the same, the right to the use of the necessary water to irrigate the unreclaimed lands shall thereupon lapse, and any subsequent right to water necessary for the cultivation of said unreclaimed land shall be acquired only by priority of utilization.[10]

Sec. 10. That it shall be lawful for any person entitled to acquire a homestead from the public lands as designated in section one of this act to settle on an irrigation farm contiguous to any irrigation district after such district has been organized by making the notifications and declaration provided for in section five of this act, and by notifying the recorder of such irrigation district, and also by complying with the rules and regulations of such district; and such person may thereupon become a member of the district and enti-

---

[8] Cinnabar is a parent mineral of mercury.

[9] Powell believed that water rights should be inseparable from the land they served; otherwise, they might be concentrated in monopolistic holdings and the associated land rendered useless for the support of farm families. Powell devotes the penultimate section of the present chapter to a further development of this argument. Generally speaking, however, his idea has not prevailed. Water rights are today held separately from land ownership and may be bought and sold, subject to restrictions imposed by the governments of the respective states. In most states, a city, industry, golf course, or farmer may purchase water rights from a farm within the same watershed and, after "retiring" the farm (i.e., taking it out of agricultural production), put the water to use in a new location and for a new purpose.

[10] The concepts of tenure by use ("use it or lose it") and of priority of utilization (seniority of right) have become cornerstones of western water law.

tled to the same privileges as the other members thereof;[11] and it shall be the duty of the recorder of the irrigation district to notify the register and receiver of the land district, and also the Surveyor-General of the United States, that such claim has been made; and such person may obtain a patent to the same under the conditions and by conforming to the methods pre-scribed in this act: *Provided,* That the water necessary for the irrigation of such farm can be taken without injury to the rights of any person who shall have entered an irrigation farm in such district: *And provided further,* That the right to the water necessary to the redemption of such irrigation farm shall inhere in the land from the time when said person becomes a member of said district, and in all subsequent conveyances the right to the water shall pass with the title to the land; but if, after the lapse of five years from the date of said notifications and declaration, the owner of said irrigation farm shall have failed to irrigate the whole or any part of the same, the right to the use of the necessary water to irrigate the unreclaimed lands shall thereupon lapse, and any subsequent right to the water necessary for the cultivation of the said unreclaimed land shall be acquired only by priority of utilization.

A Bill to Authorize the Organization of Pasturage Districts by Homestead Settlements on the Public Lands Which Are of Value for Pasturage Purposes Only

Be it enacted by the Senate and House of Representatives of the United States of America in Congress assembled, That it shall be lawful for any nine or more persons who may be entitled to acquire a homestead from the public lands, as provided for in section twenty-two hundred and eighty-nine to twenty-three hundred and seventeen, inclusive, of the Revised Statutes of

---

[11]This section makes clear that irrigation districts may not be exclusive. After initial colo-nization, new settlers may attach to a district, subject to the limitation of land and resources. They are bound by the same restrictions as the original settlers, as set forth in the remainder of this section. Such a communitarian approach to settlement ran counter to the prevailing spirit of the day, which was extremely individualistic and achieved fullest expression in the doctrine of social Darwinism, addressed in more detail in Part VII.

the United States, to settle a pasturage district and to acquire titles to pasturage lands under the limitations and conditions hereinafter provided.

Sec. 2. That it shall be lawful for the persons mentioned in section one of this act to organize a pasturage district in accordance with a form and general regulations to be prescribed by the Commissioner of the General Land Office, which shall provide for a recorder; and said persons may make such by-laws, not in conflict with said regulations, as they may deem wise for the use of waters in such district for irrigation or other purposes, and for the pasturage of the lands severally or conjointly; but the same must accord with the provisions of this act.

Sec. 3. That all lands in those portions of the United States where irrigation is necessary to agriculture shall be, for the purposes set forth in this act, classed as pasturage lands, excepting all tracts of land of not less than three hundred and twenty acres which can be redeemed by irrigation, and where there is sufficient accessible water for such purpose not otherwise utilized or lawfully claimed, and all lands bearing timber of commercial value.[12]

Sec. 4. That it shall be lawful for the requisite number of persons, as designated in section one of this act, to select from the public lands designated as pasturage lands in section three of this act, for the purpose of settling thereon, an amount of land not exceeding two thousand five hundred and sixty acres to each person; but the lands thus selected by the persons desiring to organize a pasturage district shall be in one continuous tract, and the same shall be subdivided as the regulations and by-laws of the pasturage district shall prescribe: *Provided,* That no one person shall be entitled to more than two thousand five hundred and sixty acres, and this may be in one continuous body, or it may be in two parcels, one for irrigation, the other for pasturage purposes; but the parcel for irrigation shall not exceed twenty acres: *And provided further,* That no tract or tracts of land selected

---

[12] Here Powell proposes another enormous task of classification, even greater than the irrigation survey he called for in Section 3 of the previous bill. Inventory of timber and grazing lands was never undertaken as set forth here, but it has been accomplished in an indirect and very approximate way by the establishment of federal land management agencies and the classification of lands under their control.

for any one person shall be entitled to a greater amount of water for irrigating purposes than that sufficient for the reclamation and cultivation of twenty acres of land; nor shall the tract be selected in such a manner along a stream as to monopolize a greater amount.

Sec. 5.[13] That whenever such pasturage district shall be organized, the recorder of such district shall notify the register and receiver of the land district in which such pasturage district is situate, and also the Surveyor-General of the United States, that such pasturage district has been organized; and each member of the organization of said district shall file a declaration with the register and receiver of said land district that he has settled upon a tract of land within such pasturage district, not exceeding the prescribed amount, with the intention of residing thereon and obtaining a title thereto under the provisions of this act.

Sec. 6. That if within three years after the organization of the pasturage district the claimants therein, in their organized capacity, shall apply for a survey of said district to the Surveyor-General of the United States, he shall cause a proper survey to be made, together with a plat of the same; and on this plat each tract or parcel of land into which the district is divided shall be numbered, and the measure of every angle, the length of every line in the boundaries thereof, and the number of acres in each tract or parcel, shall be inscribed thereon, and the name of the district shall appear on the plat in full; and this plat and the field-notes of such survey shall be submitted to the Surveyor-General of the United States; and it shall be the duty of that officer to examine the plat and notes therewith and prove the accuracy of the survey in such manner as the Commissioner of the General Land Office may prescribe; and if it shall appear after such examination and proving that correct surveys have been made, and that the several tracts claimed are within the provisions of this act, he shall certify the same to the register of the land district, and shall furnish to the said register of the land district, and to the recorder of the pasturage district and to the recorder or clerk of the county in

---

[13] This section is mislabeled in both the first and second editions of the report as a duplicate Section 6.

which the pasturage district is situate, and to the Commissioner of the General Land Office, a copy thereof to each; and the original shall be retained in the office of the Surveyor-General of the United States for preservation.

Sec. 7. That each person applying for the benefits of this act shall, in addition to compliance therewith, conform to the methods provided for the acquirement of a homestead in sections twenty-two hundred and eighty-nine to twenty-three hundred and seventeen, inclusive, of the Revised Statutes of the United States,[14] so far as they are applicable and consistent with this act, and may thereupon obtain a patent: *Provided,* That no person shall obtain a patent under this act to any coal lands, town sites, or tracts of public lands on which towns may have been built, or to any mine of gold, silver, cinnabar, copper, or other mineral for the sale or disposal of which provision has been made by law.

Sec. 8. That the lands patented under the provisions of this act shall be described as pasturage farms, and designated by the number of the tract or parcel and the name of the pasturage district.

Sec. 9. That the right to the water necessary to the redemption of an irrigation tract of a pasturage farm shall inhere in the land from the time of the organization of the pasturage district, and in all subsequent conveyances the right to the water shall pass with the title to the tract; but if after a lapse of five years from the date of the organization of the pasturage district the owner of any pasturage farm shall have failed to irrigate the whole or any part of the irrigable tract the right to the use of the necessary water to irrigate the unreclaimed land shall thereupon lapse, and any subsequent right to water necessary for the cultivation of such unreclaimed land shall be acquired only by priority of utilization.

Sec. 10. That it shall be lawful for any person entitled to acquire a homestead from the public lands designated in section one of this act to settle on a pasturage farm contiguous to any pasturage district after such district has been organized, by making the notifications and declaration provided for in

---

[14] That is, the applicant must furnish proof of occupancy and improvement of the land and pay a small per-acre fee.

section five of this act, and by notifying the recorder of such pasturage district, and also by complying with the rules and regulations of such district; and such person may thereupon become a member of the district and entitled to the same privileges as the other members thereof;[15] and it shall be the duty of the recorder of the pasturage district to notify the register and receiver of the land district, and also the Surveyor-General of the United States, that such claim has been made; and such person may obtain a patent to the same under the conditions and by conforming to the methods prescribed in this act: *Provided,* That the water necessary for such farm can be taken without injury to the rights of any person who shall have entered a pasturage farm in such district: *And provided further,* That the right to the water necessary to the redemption of the irrigable tract of such pasturage farm shall inhere in the land from the time when said person becomes a member of said district, and in all subsequent conveyances the right to the water shall pass with the title to the land; but if, after the lapse of five years from the date of such notifications and declaration, the owner of said irrigable tract shall have failed to irrigate the whole or any part of the same, the right to the use of the necessary water to irrigate the unreclaimed land shall thereupon lapse, and any subsequent right to the water necessary to the cultivation of the said unreclaimed land shall be acquired only by priority of utilization.

The provisions in the submitted bills by which the settlers themselves may parcel their lands may need further comment and elucidation. If the whole of the Arid Region was yet unsettled, it might be wise for the Government to undertake the parceling of the lands and employ skilled engineers to do the work, whose duties could then be performed in advance of settlement. It is manifest that this work cannot be properly performed under the contract system; it would be necessary to employ persons of skill and judgment under a salary system. The mining industries which have sprung up in the country since the discovery of gold on the Pacific coast, in

---

[15] Like the irrigation districts, a pasturage district may not exclude adjacent settlers with a legitimate interest in joining.

1849, have stimulated immigration, so that settlements are scattered throughout the Arid Region; mining towns have sprung up on the flanks of almost every great range of mountains, and adjacent valleys have been occupied by persons desiring to engage in agriculture. Many of the lands surveyed along the minor streams have been entered, and the titles to these lands are in the hands of actual settlers. Many pasturage farms, or ranches, as they are called locally, have been established throughout the country. These remarks are true of every state and territory in the Arid Region. In the main these ranches or pasturage farms are on Government land, and the settlers are squatters, and some are not expecting to make permanent homes. Many other persons have engaged in pasturage enterprises without having made fixed residences, but move about from place to place with their herds. It is now too late for the Government to parcel the pasturage lands in advance of the wants of settlers in the most available way, so as to closely group residences and give water privileges to the several farms. Many of the settlers are actually on the ground, and are clamoring for some means by which they can obtain titles to pasturage farms of an extent adequate to their wants, and the tens of thousands of individual interests would make the problem a difficult one for the officers of the Government to solve. A system less arbitrary than that of the rectangular surveys now in vogue, and requiring unbiased judgment, overlooking the interests of single individuals and considering only the interests of the greatest number, would meet with local opposition. The surveyors themselves would be placed under many temptations, and would be accused—sometimes rightfully perhaps, sometimes unjustly—of favoritism and corruption, and the service would be subject to the false charges of disappointed men on the one hand, and to truthful charges against corrupt men on the other. In many ways it would be surrounded with difficulties and fall into disrepute.

Under these circumstances it is believed that it is best to permit the people to divide their lands for themselves—not in a way by which each man may take what he pleases for himself, but by providing methods by which these settlers may organize and mutually protect each other from the rapacity of individuals. The lands, as lands, are of but slight value, as they cannot be used for ordinary agricultural purposes, *i.e.,* the cultivation of crops; but their value consists in the scant grasses which they spontaneously produce, and these values can be made available only by the use of the waters necessary for the subsistence of stock, and that necessary for the small amount of irrigable land which should be attached to the several pasturage farms. Thus, practically, all values inhere in the water, and an equitable division of the waters can be made only by a wise system of parceling the lands; and the people in organized bodies can well be trusted with this right, while individuals could not thus be trusted. These considerations have led to the plan suggested in the bill submitted for the organization of pasturage districts.[16]

In like manner, in the bill designed for the purpose of suggesting a plan for the organization of irrigation districts, the same principle is involved, *viz,* that of permitting the settlers themselves to subdivide the lands into such tracts as they may desire.

The lands along the streams are not valuable for agricultural purposes in continuous bodies or squares, but only in irrigable tracts governed by the levels of the meandering canals which carry the water for irrigation, and it would be greatly to the advantage of every such district if the land could be divided into parcels, governed solely by the conditions under which the water could be distributed over them; and such parceling cannot be properly done prior to the occupancy of the lands, but can only be made *pari passu* with the adoption of a system of canals; and the people settling on these lands should be allowed the privilege of dividing the lands into such tracts as may be most available for such purposes, and they should not be hampered with the present arbitrary system of dividing the lands into rectangular tracts.

Those who are acquainted with the history of the land system of the eastern states, and know the difficulty of properly identifying or determin-

---

[16] Powell argues that his plan is adaptable to present conditions. Had it been implemented, its execution would have depended to a great degree on the accuracy of his assessment of human nature. Powell believed the plan would work not because men and women could be trusted to do the right thing but because, acting democratically as a group, they would check each other's inclination to do wrong.

ing the boundaries of many of the parcels or tracts of land into which the country is divided, and who appreciate the cumbrous method of describing such lands by metes and bounds in conveyances, may at first thought object to the plan of parceling lands into irregular tracts. They may fear that if the system of parceling the lands into townships and sections, and describing the same in conveyances by reference to certain great initial points in the surveys of the lands, is abandoned, it will lead to the uncertainties and difficulties that belonged to the old system. But the evils of that system did not belong to the shape into which the lands were divided. The lands were often not definitely and accurately parceled; actual boundary lines were not fixed on the ground and accurate plats were not made, and the description of the boundary lines was usually vague and uncertain. It matters not what the shape of tracts or parcels may be; if these parcels are accurately defined by surveys on the ground and plotted for record, none of these uncertainties will arise, and if these tracts or parcels are lettered or numbered on the plats, they may be very easily described in conveyances without entering into a long and tedious description of metes and bounds.

In most of our western towns and cities lots are accurately surveyed and plotted and described by number of lot, number of block, etc., etc., and such a simple method should be used in conveying the pasturage lands. While the system of parceling and conveying by section, township, range, etc., was a very great improvement on the system which previously existed, the much more simple method used in most of our cities and towns would be a still further improvement.

The title to no tract of land should be conveyed from the Government to the individual until the proper survey of the same is made and the plat prepared for record. With this precaution, which the Government already invariably takes in disposing of its lands, no fear of uncertainty of identification need be entertained.

Water Rights

In each of the suggested bills there is a clause providing that, with certain restrictions, the right to the water necessary to irrigate any tract of land shall inhere in the land itself from the date of the organization of the district. The object of this is to give settlers on pasturage or irrigation farms the assurance that their lands shall not be made worthless by taking away the water to other lands by persons settling subsequently in adjacent portions of the country. The men of small means who under the theory of the bill are to receive its benefits will need a few years in which to construct the necessary waterways and bring their lands under cultivation. On the other hand, they should not be permitted to acquire rights to water without using the same. The construction of the waterways necessary to actual irrigation by the land owners may be considered as a sufficient guarantee that the waters will subsequently be used.

The general subject of water rights is one of great importance. In many places in the Arid Region irrigation companies are organized who obtain vested rights in the waters they control, and consequently the rights to such waters do not inhere in any particular tracts of land.

When the area to which it is possible to take the water of any given stream is much greater than the stream is competent to serve, if the land titles and water rights are severed, the owner of any tract of land is at the mercy of the owner of the water right. In general, the lands greatly exceed the capacities of the streams. Thus the lands have no value without water. If the water rights fall into the hands of irrigating companies and the lands into the hands of individual farmers, the farmers then will be dependent upon the stock companies, and eventually the monopoly of water rights will be an intolerable burden to the people.

The magnitude of the interests involved must not be overlooked. All the present and future agriculture of more than four-tenths of the area of the United States is dependent upon irrigation, and practically all values for agricultural industries inhere, not in the lands but in the water. Monopoly of land need not be feared. The question for legislators to solve is to devise some practical means by which water rights may be distributed among individual farmers and water monopolies prevented.

The pioneers in the "new countries" in the United States have invariably been characterized by enterprise and industry and an intense desire for the speedy development of their new homes. These characteristics are no whit less prominent in the Rocky Mountain Region than in the earlier "new countries"; but they are even more apparent. The hardy pioneers engage in a multiplicity of industrial enterprises surprising to the people of long established habits and institutions. Under the impetus of this spirit irrigation companies are organized and capital invested in irrigating canals, and but little heed is given to philosophic considerations of political economy or to the ultimate condition of affairs in which their present enterprises will result. The pioneer is fully engaged in the present with its hopes of immediate remuneration for labor. The present development of the country fully occupies him. For this reason every effort put forth to increase the area of the agricultural land by irrigation is welcomed. Every man who turns his attention to this department of industry is considered a public benefactor. But if in the eagerness for present development a land and water system shall grow up in which the practical control of agriculture shall fall into the hands of water companies, evils will result therefrom that generations may not be able to correct, and the very men who are now lauded as benefactors to the country will, in the ungovernable reaction which is sure to come, be denounced as oppressors of the people.

The right to use water should inhere in the land to be irrigated, and water rights should go with land titles.

Those unacquainted with the industrial institutions of the far west, involving the use of lands and waters, may without careful thought suppose that the long recognized principles of the common law are sufficient to prevent the severance of land and water rights; but other practices are obtaining which have, or eventually will have, all the force of common law, because the necessities of the country require the change, and these practices are obtaining the color of right from state and territorial legislation, and to some extent by national legislation. In all that country the natural channels of the streams cannot be made to govern water rights without great injury to its agricultural and mining industries. For the great purposes of irrigation and hydraulic mining the water has no value in its natural channel. In general the water cannot be used for irrigation on the lands immediately contiguous to the streams—*i.e.,* the flood plains or bottom valleys—for reasons more fully explained in a subsequent chapter. The waters must be taken to a greater or less extent on the bench lands to be used in irrigation. All the waters of all the arid lands will eventually be taken from their natural channels, and they can be utilized only to the extent to which they are thus removed, and water rights must of necessity be severed from the natural channels.[17] There is another important factor to be considered. The water when used in irrigation is absorbed by the soil and reëvaporated to the heavens. It cannot be taken from its natural channel, used, and returned. Again, the water cannot in general be properly utilized in irrigation by requiring it to be taken from its natural channel within the limits ordinarily included in a single ownership. In order to conduct the water on the higher bench lands where it is to be used in irrigation, it is necessary to go up the stream until a level is reached from which the waters will flow to the lands to be redeemed. The exceptions to this are so small that the statement scarcely needs qualification. Thus, to use the water it must be diverted from its natural course often miles or scores of miles from where it is to be used.

The ancient principles of common law applying to the use of natural streams, so wise and equitable in a humid region, would, if applied to the Arid Region, practically prohibit the growth of its most important industries. Thus it is that a custom is springing up in the Arid Region which may or may not have color of authority in statutory or common law; on this I do not wish to express an opinion; but certain it is that water rights are practi-

---

[17] In his eloquent biography of Powell, *Beyond the Hundredth Meridian,* Wallace Stegner portrayed the Major as a kind of environmental hero, a status that in many ways he deserves, but not universally. Where the existence of free-flowing streams is concerned, Powell was conspicuously unaware or unconcerned with the values of aquatic and riparian (i.e., streamside) communities, as this passage clearly shows. Powell also seems to have had little appreciation of the connection between surface and subsurface waters—i.e., that by drying up a stream one may cause a significant diminishment in the flow of springs and rivers farther down the watershed, even where no surface connection exists.

cally being severed from the natural channels of the streams; and this must be done.[18] In the change, it is to be feared that water rights will in many cases be separated from all land rights as the system is now forming. If this fear is not groundless, to the extent that such a separation is secured, water will become a property independent of the land, and this property will be gradually absorbed by a few. Monopolies of water will be secured, and the whole agriculture of the country will be tributary thereto—a condition of affairs which an American citizen having in view the interests of the largest number of people cannot contemplate with favor.

Practically, in that country the right to water is acquired by priority of utilization, and this is as it should be from the necessities of the country. But two important qualifications are needed. The *user right* should attach to the *land* where used, not to the individual or company constructing the canals by which it is used. The right to the water should inhere in the land where it is used; the priority of usage should secure the right. But this needs some slight modification. A farmer settling on a small tract, to be redeemed by irrigation, should be given a reasonable length of time in which to secure his water right by utilization, that he may secure it by his own labor, either directly by constructing the waterways himself, or indirectly by coöperating with his neighbors in constructing systems of waterways. Without this provision there is little inducement for poor men to commence farming operations, and men of ready capital only will engage in such enterprises.

The tentative bills submitted have been drawn on the theory thus briefly enunciated.

If there be any doubt of the ultimate legality of the practices of the people in the arid country relating to water and land rights, all such doubts should be speedily quieted through the enactment of appropriate laws by the national legislature. Perhaps an amplification by the courts of what has been designated as the *natural right* to the use of water may be made to cover the

---

[18] Powell is referring to the doctrine of riparian rights, under which a landowner whose land touches a flowing river or stream has a right to the continued enjoyment of that flow. As Powell predicted, riparian rights in the western United States have generally been subordinated to diversionary rights of irrigation and use.

practices now obtaining; but it hardly seems wise to imperil interests so great by intrusting them to the possibility of some future court made law.

The Lands Should Be Classified

Such a system of disposing of the public lands in the Arid Region will necessitate an authoritative classification of the same. The largest amount of land that it is possible to redeem by irrigation, excepting those tracts watered by lone springs, brooks, and the small branches, should be classed as irrigable lands, to give the greatest possible development to this industry. The limit of the timber lands should be clearly defined, to prevent the fraudulent acquirement of these lands as pasturage lands. The irrigable and timber lands are of small extent, and their boundaries can easily be fixed. All of the lands falling without these boundaries would be relegated to the greater class designated as pasturage lands. It is true that all such lands will not be of value for pasturage purposes, but in general it would be difficult to draw a line between absolutely desert lands and pasturage lands, and no practical purposes would be subserved thereby. Fix the boundaries of the timber lands that they may be acquired by proper methods; fix the boundaries of the irrigable lands that they may also be acquired by proper methods, and then permit the remaining lands to be acquired by settlers as pasturage lands, to the extent that they may be made available, and there will be no fear of settlers encroaching on the desert or valueless lands.

Heretofore we have been considering only three great classes of lands—namely, irrigable, timber, and pasturage lands, although practically and under the laws there are two other classes of lands to be recognized—namely, mineral lands, *i.e.,* lands bearing lodes or placers of gold, silver, cinnabar, etc., and coal lands. Under the law these lands are made special. Mineral lands are withheld from general sale, and titles to the mines are acquired by the investment of labor and capital to an amount specified in the law. Coal lands are sold for $20 per acre. The mineral lands proper, though widely scattered, are of small extent. Where the mines are lodes, the lands lie along the mountains, and are to a greater or less extent valueless for all other purposes. Where the mines are placers, they may also be agricultural lands, but their extent is very limited. To withhold these lands from purchase and settlement as irrigable, timber, and pasturage lands will in no material way affect the interests of the industries connected with the last mentioned lands. The General Government cannot reasonably engage in the research necessary to determine the mineral lands, but this is practically done by the miners themselves. Thousands of hardy, skillful men are vigorously engaged in this work, and as mines are discovered mining districts are organized, and on the proper representation of these interested parties the mineral lands are withheld from general sale by the Land Department. Thus, proper provision is already made for this branch of the work of classification.

In many parts of the Arid Region there are extensive deposits of coal. These coal fields are inexhaustible by any population which the country can support for any length of time that human prevision can contemplate. To withhold from general settlement the entire area of the workable coal fields would be absurd. Only a small fraction will be needed for the next century. Only those lands should be classed as coal lands that contain beds of coal easily accessible, and where there is a possibility of their being used as such within the next generation or two. To designate or set apart these lands will require the highest geological skill; a thorough geological survey is necessary.[19]

In providing for a general classification of the lands of the Arid Region, it will, then, be necessary to recognize the following classes, namely: mineral lands, coal lands, irrigable lands, timber lands, and pasturage lands. The mineral lands are practically classified by the miners themselves, and for this no further legal provision is necessary. The coal lands must be determined by geological survey. The work of determining the areas which should be relegated to the other classes—namely, irrigable, timber, and pasturage lands—will be comparatively inexpensive.

---

[19] Powell had no doubt as to the identity of the ideal man to head the survey of coal lands, irrigable lands, and timber lands: himself.

PART V

The Nation's Expert

THE ARID LANDS REPORT added luster to Powell's already substantial reputation. It established him as one of the nation's foremost authorities on the character of its vast western holdings and on the issues that attended administration of that empire. Nevertheless, the role Powell might play in addressing those issues remained unclear, for even as the *Arid Lands* report found its first readers, Congress again began to examine the question of rivalry among the western surveys. Only three remained in the field, the members of Clarence King's survey of the fortieth parallel having turned to writing reports and publishing their findings. Congress, however, did not resolve the question speedily. The politics of the matter being delicate, it asked the National Academy of Sciences to examine the matter and recommend a solution.

Powell had ample opportunity to explain his views to members of the committee the academy subsequently appointed, among whom he recognized many friends. He advised that all government survey work be consolidated into a single operation under civilian control, which should devote its principal energies, he said, to geographical and geological researches, as opposed to military or economic questions. In view of the rapid changes endangering the lifeways of Indian tribes, he further argued that the new survey should also pursue ethnographic work and record the quickly vanishing languages, traditions, and practices of the West's native people. In short, Powell advised that the consolidated survey conduct its business as he had been conducting his own.

The academy largely agreed with Powell's ideas, and to a degree, so did Congress, which passed legislation to consolidate the surveys under civilian leadership in February 1879. On March 3, the same day that Congress ordered a second printing of the *Report on the Arid Lands,* President Hayes signed the act creating the United States Geological Survey and appointed Clarence King its director. Powell was not a serious candidate for the job and received a lesser plum. In keeping with the urgings of the *Arid Lands* report, he had argued for reform of the survey methods used in parceling the public lands, a course of action the new legislation rejected. The legislation also failed to instruct the survey to conduct ethnographic research. Instead, Congress created the Bureau of Ethnology within the Smithsonian Institution and made Powell its founding director.

This was to Powell's liking, for anthropological and ethnographic questions increasingly held the strongest claim on his personal research interests, and that claim did not weaken with time: He remained director of the bureau until his death and in the course of his tenure sponsored scores of vital studies and nurtured an entire generation of anthropologists. Nevertheless, his involvement in land administration and survey issues hardly abated, and Powell soon found himself juggling multiple roles. In July 1879, he was appointed to a newly created Commission to Codify the Land Laws, and for the balance of the year he traveled thousands of miles throughout the West, visiting its most isolated corners, hearing voluminous testimony, and sharing the message of the *Arid Lands* report with the widest possible audience. Little more than a year later, in the spring of 1881, Clarence King resigned from the Geological Survey, and Powell was appointed his successor. He now had two young agencies to organize and run, two staffs to recruit and direct, two budgets to administer, and two sets of issues to advance or defend before Congress, as opportunity allowed. He also had his own research to pursue, chiefly involving the classification of Indian languages. How he found time for all this might astound the merely human among us. Powell was an engine of work and in his prime.

The warnings of the *Arid Lands* report had been aired and its proposed reforms rejected. One of Powell's most controversial proposals was his recommendation that the government grant from the public domain pasturage farms of unprecedented size: 2,560 acres, or 4 square miles. To many people such an idea seemed a capitulation not to the realities of land and climate, as Powell would have it, but to the desires of monied interests: large-scale ranchers and the investors who backed them. The belief grew that Powell's program was designed to serve the accumulation of capital by the elite and to exclude settlers of modest means from the West. And so its prospects waned.

But beginning late in 1886, nature began to have its say, and it seemed to agree with Powell. That winter, horrific blizzards swept the northern plains. Range cattle died by the hundreds of thousands, and ranchers large and small, especially those who had ignored Powell's admonition to grow winter feed on irrigated land, found themselves ruined. Then followed a prolonged drought, and homesteaders who had thronged to the Great Plains and points west during the first half of the 1880s saw their dryland farms turn to dust. Hopeful settlers became desperate refugees, withdrawing from the expanses of the arid region with their possessions piled high on overloaded wagons, much as a future generation would flee the Dust Bowl of the southern plains.

People cursed the weather, but it was the land laws of the United States that had played on them a cruel joke. The 160-acre homestead was an absurdity where rains were scant, and even its progressive enlargements in the years ahead would offer no help if they merely permitted crop failure on a larger scale. Only irrigation would enable farmers to survive the vagaries of dry-country weather.

Congress took action by dusting off one of Powell's core ideas. In 1888 it authorized an irrigation survey as a unit of the Geological Survey. The purpose of the new survey was to identify the reservoir sites and canal rights-of-way necessary to develop irrigation in the arid lands. These lands were then to be withdrawn from the public domain and reserved for the construction of an appropriate infrastructure.[1] Powell and his men set to work, first with training, then by sending crews to do the actual surveys. So far so good.

Speculators, however, kept a close eye on the crews of the Irrigation Survey. It took no great intelligence to divine which lands the new surveys would propose to irrigate, and with water the value of those lands would rapidly appreciate. So speculators laid claim to the best land wherever prospects looked good. Such interests understood that if they possessed title, or merely a defensible interest, in land essential to an irrigation system, they might later sell their stake at a profit. And so they hastened to file on land before it could be withdrawn.

These issues came to a head at Bear Lake, on the border of Idaho and Utah. In August 1889, Interior Department officials invoked an obscure amendment to the Irrigation Survey's first appropriation bill, authorizing it to withdraw from entry "all lands made susceptible of irrigation." Most observers, taking the language at face value, believed that the amendment would affect only a small subset of the public domain. But in consultation with the president and his advisors, the Department of Interior determined to apply it to the entirety of the public domain and to close all public land to entry. Moreover, it made the closure retroactive to October 2 of the previous year, throwing nearly a year's worth of claims into administrative limbo. Although the amendment included an escape clause that allowed the president to reverse the closure of selected public lands, President Harrison chose not to exercise it.

The resulting furor doomed the Irrigation Survey. Land filings in the late 1880s and early 1890s averaged 25 million acres a year.[2] Dryland farming may have faltered on the prairies, but Americans were still hungry for

---

[1] A word about the traditional vocabulary of the public domain: lands are said to be open to entry until they are withdrawn. Withdrawing lands from the public domain means reserving them for a particular use, such as an Indian reservation, a forest reserve (which we today call a National Forest), or, as in the earlier example, a federal reservoir or canal site. Lands thus withdrawn are said to be closed to entry, which means citizens may not claim them for use as mines, for home sites, or for other private purposes.

[2] Pisani, *To Reclaim a Divided West,* 139.

land, still settling and speculating, still moving west across the plains and east from California and churning the ownership and occupancy of every state and territory in between. Nothing since the Civil War and Reconstruction had made so many Americans so mad.

Although Powell's views inspired the closure of the public domain, in this instance he had urged neither the original amendment nor the Interior Department's order. Powell's innocence, however, did little to protect him or his survey from the anger of politicians and the public. Nor did he seek protection. When asked whether he supported the closure, he honestly replied that he thought it was a good idea. Pressure to reopen the public domain mounted rapidly. Pleading for more time, Powell argued that the survey, if it were to have value, must be accurate, the maps exact. His position was intensely unpopular but he stuck to it, and the storm raged on.

It reached its highest pitch in the summer of 1890 when Powell submitted his plan of operations and proposed budget to Congress. The ensuing hearings were the bitterest of his long career, but he did not approach them defensively. As he had also done earlier that year in hearings on a bill to cede public domain in the arid lands to the states, Powell gave his congressional audience what amounted to an intensive course on the realities of western lands, complete with charts and stunning maps. He argued that the government was obligated to guard against a repeat of the suffering of homesteaders in the late 1880s and that the Irrigation Survey, as time-consuming and nettlesome as it might be, was the proper way to do so. Congress did not agree. Powell asked for an appropriation of $720,000 for the portion of the Geological Survey from which the Irrigation Survey would draw its funds. Congress gave him $162,500, effectively eliminating the Irrigation Survey, and rescinded the Interior Department's power to reserve irrigable land.[3]

Congress meanwhile provided $719,400 for the more general mapping and inventory work of the Geological Survey's work, sustaining it for the moment as one of the best-supported scientific organizations in the world.

---

[3]See discussions Worster, *A River Running West,* 482–83, 501–22, in Stegner, *Beyond the Hundredth Meridian,* 328 ff., and Darrah, *Powell of the Colorado,* 310–11.

But Powell's political star had passed its zenith, and it never rose again. His enemies came after him even more vigorously the following year and slashed the budget of the Geological Survey nearly in half, forcing the elimination of 14 positions. Powell gained back some ground in the 1892–93 budget year, but the hard financial times brought on by the panic of 1893 helped foreclose any prospect that he might rebound fully. In 1894 he resigned as director, turning his attention to anthropological and philosophical matters and to his personal health. The nerves in the stump of his arm had regenerated and gave him great pain. He needed surgery again.

The following three essays should be understood in the context of the brief and pyrotechnic arc of the Irrigation Survey. In the first, "Trees on Arid Lands" (1888), Powell finds himself in the familiar position of telling people what they do not want to hear. He argues against the then popular fiction that tree planting and other activities of settlement can substantially alter the arid climate of the West. He also clearly enunciates his utilitarian view of the natural world: "Man cannot change the great laws of nature; but he can take advantage of them, and use them for his purposes."

In "The Lesson of Conemaugh" (1889), he is concerned with turning an apparent setback into an opportunity to win support for the Irrigation Survey. On May 31, 1889, a privately built earthen dam at the confluence of Stony Creek and the Conemaugh River in Pennsylvania had burst, unleashing the legendary Johnstown Flood, which killed 2,200 people. Critics of the Irrigation Survey, who included forest conservationists, said that the disaster demonstrated the inadvisability of building reservoirs of any kind. Conservationists added that where the development of irrigation was at stake, forest protection, not dam building, offered the surest and safest means to ensure a dependable water supply. Powell's reply is methodical and ponderous. He does not deliver his main message until halfway through the essay, but at last he points out that the cause of tragedy at Johnstown was not the principle of dam construction but the specific fact that the Conemaugh dam was badly designed. It failed because it was too small to withstand the predictable pressures of heavy rainfall within its watershed. The disaster illustrated not the inherent danger of dams, but the need for detailed and thorough survey work wherever dams were to be sited. Only in the context of such survey and analysis, performed by qualified experts, might a dam in all safety be properly "related to the natural conditions" and designed for "the duty the dam was required to perform." In other words, the lesson of Conemaugh was that an irrigation survey like that which Powell headed was not less important, but more so.

Powell's address to the Montana Constitutional Convention in August 1889 took place in the course of a lengthy tour of the arid region under the auspices of the Senate Special Committee on the Irrigation and Reclamation of Arid Lands. When Powell gave it, he was at the zenith of his power and influence: The Interior Department's fateful order closing the public domain was days away but had not yet plunged the Irrigation Survey into a maelstrom of controversy. Powell used the occasion to urge a novel concept: the organization of political units in conformance with geographic realities. As Powell saw it, county boundaries (or the boundaries of political subdivisions analogous to counties) should follow the natural boundaries of watersheds. This is the germ of one of Powell's most profound and revolutionary ideas. He argued it before Congress through the spring of 1890, illustrating it with maps that showed the West divided into watersheds (see maps 7 and 8, this volume). The politicians could not have helped noticing that those watersheds, in turn, dissolved the existing boundaries of states and territories, the very units from which they had been elected and to which they owed their political loyalty. Nevertheless, Powell pressed on, and he shared his bold plan with the public through the trio of articles collected in Part VI of this anthology.

Unfortunately for Powell, his urgings fell on deaf ears in Helena. A glance at a modern map of Montana shows that the Montanans, although they received Powell warmly, failed to align their counties with watersheds. Indeed, of all the embryonic western states, only Wyoming heeded one of Powell's major recommendations, which was to tie water rights inseparably to the land.

A further point of interest in the Montana address is that it appears to have been compiled from transcription (and a hasty one at that), not from a written text provided by Powell. Thus it reflects how he actually spoke. Here, if anywhere, we may gain the best indication of the native cadence and tenor of his voice.

Even after the Irrigation Survey had been dismantled, support for irrigation continued to grow, and Powell remained as true a leader as the movement had. In October 1893, the second International Irrigation Congress convened in Los Angeles, and Powell was a featured speaker. The gathered throng greeted him with enthusiasm as he approached the podium. But Powell was feeling sour. The giddiness and lack of realism in the previous speeches troubled him. Irrigation had a great future but not the unlimited one that its boosters were claiming for it. Powell put aside his prepared remarks. He used his time on the podium to remind the congress that not all the West could be reclaimed. He explained that water resources were sufficient to irrigate only a fraction of its potentially arable land. In fact, he said, most regions would be hard pressed to irrigate the arable land that had already been conveyed from the public domain. Extensive additional grants to individuals were therefore unnecessary and unwise.[4] The assembled advocates of irrigation development liked Powell's message of limits no better than the boosters of dryland farming had liked Powell's gloomy warnings a decade earlier. They jeered and booed him. Some stood up and argued with him from the floor. Once again, John Wesley Powell was telling people what they did not want to hear.

---

[4] International Irrigation Congress, *Official Proceedings,* 2: 107–16.

SELECTION 9

———◇———

Trees on Arid Lands

———

The editor of the prestigious journal Science *introduced this essay by noting that "Major J. W. Powell, director of the United States Geological Survey, has written the following interesting letter to the* Kansas City Times, *thus making a valuable contribution to the discussion of the subjects of forest-growth on arid lands, the effects of hot winds, and the extent to which irrigation may change the agricultural climate of the plains."*

———

The plains are treeless because they are arid. There is an opinion widely existing in the popular mind, and springing up in the current literature of the West, which is opposite to this, to the effect that the dryness of the climate is the result of the lack of forests. An argument in favor of tree-planting and forest-culture has often been based on this error. The effect of forests upon rainfall has been investigated by many methods, in many countries, and at many times, and the result of all this investigation shows that the presence or absence of trees influences the general rainfall or amount of precipitation only to a very limited degree. It is, in fact, not certain that their presence does increase rainfall; but it is certain, that, if it does, the increase is so slight as to play but an insignificant part as a climatic factor.[1]

———

Science 12/297 (October 12, 1888): 170–71.

———

[1] Powell is arguing against a powerful theme within a then popular school of wishful thinking (see Smith, "Rain Follows the Plow"). Smith discusses the subject in briefer terms in Chapter 16 of his groundbreaking study *Virgin Land.* The fanciful notion that activities of settlement from tree planting to railroad construction would moderate the climate of the plains helped encourage settlement of the arid lands in ways that ensured disappointment,

Yet forests, or abundant trees, exert an influence upon climate in its relations to agriculture. Two ways in which this influence is exerted are worthy of careful consideration.

First, while it is not probable that forests diminish or increase the total amount of rainfall in any country, yet it appears that forests regulate this rainfall, so that there are fewer fierce storms and more gentle rains. When the rains fall in storms, the water is speedily gathered into streams, and at once passes from the country; but, when they fall in gentle showers, time is given to moisten the soil and invigorate vegetation.

Second, forests provide against the speedy evaporation of the water by protecting the lands from the fierce rays of the sun, and more especially by protecting the land from the rapid passage of dry winds, which drink up the water from the soil and growing plants with great avidity.

It is manifest that the effect of the forests upon the great movements of the atmosphere must be very slight when due regard is given to proportions between cause and effect. Forests can affect only the winds close to the earth by creating a friction at the surface; but the soil, and the smaller plants growing therein, may be greatly sheltered by trees. Though the general climate may be scarcely affected, the agricultural climate may be materially modified.[2]

The relation of forests to humidity, and of prairies and plains to aridity, should be clearly understood. In middle latitudes, and under average conditions of relative humidity, low, gnarled forests will be produced with

---

[1](*continued*) suffering, and defeat. A striking if comical example of this delusive creed appeared in the Santa Fe *New Mexican* on June 5, 1886: "As to the theory of this improved and more equitable distribution of rain fall in this section, there no longer seems to be room for doubting that tree planting, turning over the soil, and the atmospheric shocks occasioned by the running of trains and the introduction of the telegraph, have all had a beneficial influence. The recent forest fires which have prevailed in the Santa Fe mountains may also have had some effect in drawing thither the rain clouds, but no matter what the local cause, the opinion seems to prevail that henceforth the broad valley of the Rio Santa Fe will spring into popularity as one of the choice farming districts of the west."

[2] Powell, quite rightly, is distinguishing between regional climate and the microclimate at soil level that most influences plant life.

about ten or twelve inches of rainfall; that is, in the Rocky Mountain region, and generally on the Great Plains, forests of cedar and piñon[3] can be produced with a little more than ten inches of rain annually. Now, this is a well-established fact. Why, then, are the arid valleys and Great Plains treeless? The answer is, that the fires destroy the trees, and prevent their growth. In a region of great humidity, say, of forty inches or more of rainfall, forests are largely protected from fire by such general humidity. In regions of country having from ten to twenty-five inches of rainfall, all forests are destroyed unless protected by art or topographic position. In regions having between twenty-five and forty inches of rainfall, prairie-lands interspersed with timber-lands will usually be found; that is, in ordinary seasons, trees will be protected from destructive fires by the general humidity, but in excessively dry seasons the trees will be destroyed, now here, now there; so that, by the natural process of tree-propagation, the forests will encroach on the prairies, and through the fires of excessively dry seasons the prairies will encroach on the forests; and so prairie conditions and forest conditions forever contend with each other for the possession of a sub-humid land.[4] In the direction in which aridity increases, prairie conditions will more and more prevail; and, as humidity increases, the forest condition will more and more prevail. In general it may be stated, that, other things being equal, the dryer the climate, the smaller the forests; the wetter the climate, the greater the forests; for, although the rainfall may be sufficient to grow forests, it may not be sufficient to protect them from fires. The Great Plains and the valleys of the Rocky Mountains are all capable of sustaining forests of certain trees adapted to the climatic conditions found therein.

It is possible, and in due time it will be practicable, for man to clothe the

---

[3] Wallace Stegner points out that the West is a land of misnomers: The prairie dog is not a dog; the horned toad is not a toad but a lizard; the jackrabbit is a hare, the pronghorn antelope a goat, the buffalo a bison, etc. (See "Living Dry" in Stegner, *Where the Bluebird Sings to the Lemonade Springs.*) And, as Powell probably knew, the cedar is not a cedar, but a juniper, of which about eight species (depending on who's counting) may be found in the West. The piñon is *Pinus edulis.*

[4] This is a balanced, clear description of an ecological relationship that few Americans in Powell's time appreciated.

naked lands of the Great Plains and the arid valleys of the West with forests without artificial irrigation.[5] From this must be excepted certain desert-lands west of the lower portions of the Colorado River, where the rainfall is insufficient, and also certain tracts of bad-lands which will always be tree-less for reasons that need not here be described.

The amount of rainfall necessary to produce forests in any given latitude will depend to some extent upon the character and conditions of the soil, some soils needing more rain than others for this purpose; but the soil con-dition has narrow limits.

If it be true, as has been asserted above, that the arid plains and valleys may all produce forests without artificial irrigation if protected from fires, how are such forests to be planted, in what manner can they be protected, and how shall the trees adapted to the climate be selected? These are the practical problems to be solved.

Great areas of uninhabited land cannot be redeemed and protected; the protection must come from men living on the land, and utilizing it for agri-cultural and pasturage purposes. The way in which this can and will be done may be briefly and crudely sketched as follows:

Adown the valleys and across the plains flow many streams of water—brooks, creeks, and rivers—that have their sources in the mountains by which the arid lands are dissected; and all of these streams can be utilized to irrigate the dry and parched lands that now present the desolation of deserts. By their use many tracts of land scattered far and wide throughout the whole country may be brought under cultivation, and covered with growing fields and luxuriant groves. In this manner populous and prosper-ous settlements may be distributed throughout that land of drying winds and scorching suns. When industrious and thrifty people once get a foothold in this manner, they plant orchards and vineyards, and surround their farms and fields with trees, and plant them by the roadside, and every man devotes a part of his farm to timber-culture, and the naked lands are speedily covered with a rich vegetation. A generation ago the prairie region

---

5 Through tree planting and fire suppression.

east of the Missouri River was so destitute of forests that large districts were supposed to be practically uninhabitable; but since that time it has been covered with orchards, vineyards, and groves, and now, from the lands that were once so naked, millions of trees spread their branches to the breezes. In the same manner, by means of artificial irrigation, great numbers of tracts of land will be cultivated throughout the arid country, and diversified groves will be developed. But not all the arid lands can be redeemed, as the water of all the living streams is inadequate to the task; but the intervening land will be utilized for pasturage purposes, and will be protected by the people from fire, and groves will be planted, and the face of the country not under cultivation will be forested.

In the region practically uninhabited the water now flows from the mountains to the sea; but, when the streams are utilized in irrigation, the water will be evaporated, and the humidity of the climate will be increased thereby, and dry winds will no longer desiccate the soil and shrivel the vegetation. As the general humidity is increased, the moister air, as it drifts eastward in great atmospheric currents, will discharge more copious rains, and the humid region will extend farther westward, and the arid region will correspondingly shrink in its proportions.[6] Irrigation will increase the humidity of the climate, and increase protection from fires to the non-irrigated lands; and, as the lands gain more and more water from the heavens by rains, they will need less and less water from canals and reservoirs. When all the water of the arid country is ultimately appropriated for irrigation by using all the streams through the season of irrigation, and by storing the surplus that flows through the non growing season,[7] and by collecting in the reservoirs the storm-waters of the streamless valleys, the general humidity of the atmosphere in the arid region will be increased, and hence the rains will be increased, and a smaller amount of artificial irrigation will be

---

[6] This is Powell's greatest concession to the argument that human activities will change the climate of the arid lands. Although irrigation does increase humidity locally, such a prophecy did not come to pass over large areas to any measurable degree.

[7] Compare with Powell's advice to delegates to the Montana Constitutional Convention (Selection 11). Powell consistently urged the full exploitation of water resources.

needed. By all of these means a large share of the arid lands will be redeemed. But all will not be redeemed: there will still be extensive areas of pasturage-lands not under the plough, for all that man may do will be insufficient to radically change the climate. The non-irrigated lands can be greatly improved by extensive tree-planting; but as these trees are to be supported by the general rainfall, which is scant, it will be necessary to select trees adapted to arid conditions, and this will require extensive experimentation. The wide distribution of the cedar, and the piñon or nut-pine, throughout the country under consideration, points out the fact that these two trees may be widely used; but there are many others on the Pacific coast which perhaps will be more valuable; and it will probably be found that there are many trees in the arid lands of the eastern hemisphere which can be introduced with advantage.[8] But this tree-planting is a question of a somewhat remote future. At present the trees planted in the arid region will depend for their existence and vigorous growth upon irrigation, and the experiments demanded at the present time must be with such trees.

The great currents of air which now traverse the plains are impelled by agencies that produce the general circulation of the atmosphere throughout the globe, modified by the general configuration of the plains in its relation to the mountains of the West and the low humid lands of the East. These general conditions cannot be modified by man; and the storms will come and the winds will blow for ages as they now do, unchanged by the puny efforts of mankind; and yet the agricultural conditions of the country may be greatly modified and improved by the efforts of man. Man cannot change the great laws of nature; but he can take advantage of them, and use them for his purposes.

There is a theory held by some persons in the West that rainfall is largely dependent upon the electrical conditions of the atmosphere, and that these conditions are modified by the various changes wrought by the hand of

---

[8]Siberian elm, tamarisk, and Russian olive are among the Eurasian trees that are today ubiquitous throughout the arid lands of the West.

man in the settlement of the Great Plains.[9] As this appeal is to some occult agency, it becomes quite popular to those who love to revel in the mysteries of nature. Of course, it is never explained. It is a case where cause and effect are confounded. Atmospheric electricity is the result of certain conditions and movements in the atmosphere. To explain atmospheric changes by attributing them to electricity is like explaining the origin of the fire by the light it produces, or like explaining the explosion of the powder in the cannon by attributing it to the roar which may be heard in the distance. The electricity in the air is related to atmospheric changes as effect is related to cause.

In conclusion let it be said, first, that a large body of the arid lands can be redeemed by irrigation, and that the agriculture resulting therefrom will be in the future, as it has ever been in the past, the highest condition of agriculture, for the agriculture which is dependent upon rains is subject to storms on the one hand, and to droughts upon the other; but, when the water-supply is properly controlled by the arts of man, the soil is made to yield its most abundant returns; second, that, under the culture and protection of man, vineyards, orchards, and groves can be established over vast areas, where, under the control of nature, only deserts are found; third, the siroccos[10] of the Great Plains cannot be tamed, but men may protect their homes, their gardens, and their fields from devastation by them; fourth, the lightnings of heaven cannot be employed to bring rain upon the plains, but electricity may be used to illumine the cities and towns and hamlets that must ultimately spring up over all that land.

---

[9]See note 1, page 219 this section.

[10]In original usage, a hot, dry, dusty wind blowing from the Libyan deserts to Sicily, Malta, and southern Italy.

The Lesson of Conemaugh

The horrors of the 1889 Johnstown Flood intensified public mistrust of all kinds of dam construction, and opponents of reclamation heightened their criticism of policies that would stimulate construction of irrigation reservoirs. Here is Powell's response.

The experiences of civilization teach many lessons that go unheeded until some great disaster comes as an object-lesson to recall to men's minds things known but half-forgotten. The Conemaugh disaster belongs to this category. For more than four thousand years civilized men have been constructing reservoirs in which to store water for various purposes. The conditions to be fulfilled in their construction are well known, for the lesson has been enforced upon mankind from the dawn of civilization to the present time by disasters too many to be enumerated.

Hydraulic engineering is the oldest scientific art. No other can compare with it in this respect, except that of architecture in its application to the building of temples and pyramids; but scientific engineering is even older than scientific architecture. Everywhere throughout the world civilization began in arid lands, and hydraulic engineering was the first great problem to be solved; and for this reason it was solved at an early time, and well solved. Something has been added through the years, but not much. In our own times these problems have come to be of far greater importance than they were in antiquity, and the civilized world has now reached the dawn of

---

North American Review 149 (1889): 150–56.

a day of hydraulic engineering of such magnitude that all the works hitherto accomplished are insignificant compared to those now to be planned and executed. Let the significance of this statement be briefly set forth.

One of the purposes for which hydraulic engineering has been prosecuted in late civilization is the utilization of powers otherwise running to waste. For a time a check has been given to this form of development by the introduction of steam, but at present the great transition in modern industries is from the employment of muscular power to the employment of the physical powers of nature, and it is probable that the resort to water-power will rapidly increase in the immediate future. It certainly will if the dream of modern electrical science is realized, so that water-power can be economically converted into electric power and transported from place to place. If this is done—and its accomplishment is hardly to be considered Utopian— all our highland streams will immediately become of value as powers,[1] and dams and reservoirs must be constructed in far greater numbers than in the past.

Modern sanitary science condemns well water for domestic purposes in cities, towns, and villages.[2] Disease is at the bottom of a well; health in the waters of the heavens; and the people must have this pure water. The demand for highland waters for such purposes is rapidly increasing. The speedy development of city and town life under the new industrial conditions makes this one of the most important uses to which water can be applied. Wherever the houses of men are clustered reservoirs or systems of reservoirs must be built. Nothing can be more certain than that the storage of water for this purpose will greatly and quickly increase throughout the United States.

Along the course of every river there is a flood-plain of greater or less width. This is the plane established by the sediment washed from the hills and upper country and deposited along the course of the river outside of its low-water channel, but within the area covered by water at the time of great-

---

[1] That is, sources for generating power, both mechanical and electric.
[2] In densely populated areas, sewage taints the groundwater on which wells depend.

est floods. These greatest floods are infrequent, and are not coincident with the annual floods, but much higher. The plane of the ordinary flood is much lower than this great flood-plain, which is established by the maximum floods, occurring ten, twenty, or even fifty years apart. Such flood-plains are the most fertile lands, and always tempt the agriculturist. Yet torrents sweep over them from time to time, destroying property in vast amounts, and even life. In lands already highly cultivated, densely populated, and of great value, protection from floods has come to be an important problem. One, and only one, method of protection is possible: the flood waters must be stored and allowed to find their way to the sea during times of low water. The preservation of lands in this manner accomplishes another end, as the navigable streams are improved thereby. Great floods destroy low-water channels by blocking them with natural dams. By storing the water of such floods, and discharging it during low-water time, these channels are kept open and a more equable volume is preserved.

There is another use to which flood waters are put. Experience has shown that they contain vast stores of fertilizing elements. All other fertilizers that man can utilize sink into insignificance when compared with those furnished by flood waters. In highly-civilized and densely-populated lands this source of fertilization is already used, and it will be used more and more as the years go by. In the United States we are just beginning to appreciate this. The conditions under which agricultural operations have hitherto been carried on have not directed the attention of our farmers to this subject until of late years. It is far within the facts to state that any region of our country may have its agricultural production doubled by the use of its flowing waters for the fertilization of the lands.[3] The time is rapidly coming when the flood waters of the country will be used for this purpose on a grand scale, and reservoirs will be constructed all over the land, as they are now in process of building in England, Germany, France, Italy, and other countries.

_____

[3] See note 2, p. 150, of Selection 6 (Preface to *Report on Lands of the Arid Region*) concerning the problem of silt in irrigation systems.

About two-fifths of the area of the United States is so arid that agriculture is impossible without artificial irrigation, the rainfall being insufficient for the fertilization of ordinary crops. In this region all agriculture depends upon the use of running streams. In all of this country, wherever agriculture is prosecuted, dams must be constructed, and the waters spread upon the lands through the agency of canals. Again, as the season of growing crops is comparatively short—in most of the country it lasts from two to three months only—the waters of the non-irrigating season will run to waste unless they are stored in reservoirs. Already the storing of these waters is begun; the people are constructing reservoirs, and will continue the process until all of the streams of the arid region are wholly utilized in this manner, so that no waste water runs to the sea. Less than a third of the streams of the arid region run to the sea, even now, as the great majority are "lost rivers."[4] A little further explanation is necessary to understand how these waters are to be utilized.

The arid region is mountainous. Mountain ranges enclose valleys, while plains, mesas, and plateaus carry dead volcanoes on their backs. The precipitation of moisture on these lands is confined to the mountains, where it is excessive. The fertile lands along the plains and mountains are arid. In all the region agriculture is possible only by collecting the mountain waters and using them on the plains and valleys. Wherever a farm, a village, or a city is made, hydraulic works are necessary, and dams must be constructed and reservoirs built. Considering the whole country at large, its hydraulic industries are gigantic. In the region of country where land is more abundant than water, the value inheres in the water, not in the land. Land, like air, is found in greater quantities than can be used; water is the necessary, and value is given to the land by the water-right which it carries; if the water-right is dissevered, the land is valueless. These are not unfavorable conditions for agriculture, however. The farmer's industry is more attractive and

---

[4] A lost river is a river that loses its surface flow before reaching a larger body of water. Many streams in Nevada, for instance, carry mountain runoff into desert valleys where they gradually sink away and vanish into the sand and gravel of their own alluvium.

more profitable in an arid than a humid region. All of the early civilization of
the world began in arid lands, and the best agriculture of the world today is
carried on by means of artificial irrigation. The seemingly-desert plains of
the arid region of the West are, in fact, abundantly rich when watered artificially. The gentle storms of a humid region fructify the land, but its tempests
drown vegetation. In such regions the planting season is now too dry and
now too wet, and many a prospectively-rich harvest has been destroyed by
a harvest-time storm. Agriculture in arid lands is not subject to these vicissitudes. The mountains catch the floods, while the valleys are fertilized by
the hand of man, who turns the waters upon them at his will. At the day and
hour he pleases he spreads the water upon his garden, his vineyard, or his
field in quantities governed by his judgment. When harvest time comes, he
reaps his field with a mind free from fear of storms. Ultimately one of the
great agricultural regions of the country will be found in the irrigated plains
and valleys of the West. Sage-brush plains, sand-dune deserts, and alkaline
valleys will be covered by gardens, fields, and groves, all perennially fertilized from thousands of mountain lakes.

Enough has been said to show that the storage of water in reservoirs is
not one of the trivial incidents of modern industry, but one of its most
important factors, and that in this country we have only reached the beginning of its development. We may expect, in the course of a few generations,
that all the highland streams of America will be controlled and utilized, and
that the floods will be bridled and become the trained servants of man, as
wild beasts have been domesticated for his use.

In view of these facts, it is only the thoughtless man, governed by the
impulse of the hour, and dragged from the throne of his reason by the emo-
tions which arise at human suffering, who will believe that the vast industries which have been mentioned must be stopped because hydraulic
power, when improperly controlled, may become an agent of destruction.
Badly-constructed houses may fall and overwhelm families, but no check to
the construction of houses will be made thereby. Fires will cause conflagration, yet homes will be warmed. Bridges may give way and trains leap into
the abyss, yet bridges will be erected. Cars will leave the track and plunge

travelers over embankments, but railroads will be operated. Dams will give way and waters overwhelm the people of the valley below, but dams will still be built.

What lesson, then, is there in the Conemaugh disaster? Nothing new to scientific engineering, but a very old lesson, that must needs be taught to mankind again and again. From the accounts, which have appeared in the scientific journals, it seems that the dam was properly constructed. Earth dams are more common than all others. Most of the dams constructed for four thousand years have been, in all essential particulars, like that at Conemaugh. Where, then, was the trouble? In the construction of the dam there was a total neglect to consider the first and fundamental problem—the duty the dam was required to perform. The works were not properly related to the natural conditions, and so a lake was made at Conemaugh, which was for a long time a menace to the people below, and at last swept them to destruction.

When the construction of such a dam is proposed, the first thing to be done is to determine the amount of water to be controlled and the rate at which it will be delivered to the reservoir under maximum conditions of rainfall or snow-melting. The proper method of procedure is to determine, first, the area of the drainage basin supplying the reservoir; second, the declivities of the supplying basin.[5]

The very first thing, then, is a topographic survey.

The second need is a hydrographic survey.

The precipitation in rain and snow over the basin must be determined as an average from year to year, and also the maximum precipitation at the times of great flood. This must be supplemented by the gauging of streams to determine their average volume and maximum volumes. All of these factors are necessary and preliminary to the construction of a safe and efficient reservoir system by making mountain lakes. Before a reservoir dam is constructed, it is of prime importance to determine what will be required of it. With these facts ascertained, the engineer can easily plan works ade-

---

[5] That is, the area of the basin and the steepness and contour of its slopes.

quate to control the forces involved; he can readily determine how much water he can store, and what waste-way will be necessary to discharge the surplus.

The art of dam-construction is quite within the grasp of every intelligent engineer. In the case of solid masonry dams, the waste-way is over the whole surface of the dam,[6] as at the Great Falls of the Potomac, where a dam has been constructed to divert the water into the reservoirs that supply Washington. But masonry dams are few; earth dams, and those related to them, are many, and, with these, special waste-ways must be provided, adequate to meet all possible emergencies. The rules for their construction are well known, and have been known for tens of centuries. In American engineering, that which has been most neglected is a precise determination of the duty of the dam—the conditions which it must fulfill or else be destroyed. These can be determined only by a topographic survey, which gives the area of the drainage basin and its grade-curves. To this must be added a hydrographic survey, which may have to extend over some years. It is not necessary that this survey in all of its parts should be over each basin where a reservoir is to be constructed. The average and maximum rate of precipitation may be determined for large regions, and the general facts used for the several cases, always allowing a margin for safety. But the topographic survey and the stream-gauging are essential to each individual basin.

There are other factors to be determined that are important to persons engaged in constructing reservoirs for industrial purposes. Two may be mentioned here. The streams feeding the reservoir should be gauged for the purpose of determining the amount of sediment they carry, in order that the life of the reservoirs may be known, or that the proper engineering appliances may be devised to discharge such sediment; and the rate of evaporation should be ascertained, so as to know how much water is lost thereby.

---

[6] Large-scale masonry dams require carefully designed spillways, much as do the smaller earthen dams described here by Powell. Hoover and Glen Canyon dams, for instance, could not long withstand a sustained flow of water over their crests and down their massive concrete faces.

In the construction of reservoirs in the arid region there are important problems not pertaining to humid regions. To a large extent the sources of the water are in high mountains, where the chief precipitation is snow, which, to some extent, stores itself in snow-banks and glacial fields, to be melted by the summer sun at the time when irrigation is required. The upper portions of these mountains are largely treeless—a condition favorable to the storage of snow. In a forest region the snows are evenly distributed over the entire surface, and are quickly melted when the summer rains and suns come; but in a treeless region the snows are accumulated in great drifts in the lee of rocks and cliffs and under the walls of gorges and canyons. Such great drifts are themselves stupendous reservoirs of water, and artificial works are necessary only to control the flow properly and distribute the water at the places and times needed. Wherever the chief precipitation is snow, forests are a disadvantage if the waters are needed in the valleys below for irrigation, for the forests keep the snow distributed over broad areas of ground and expose it to the winds on their trunks, branches, and leaves, so that altogether the mountain evaporation is enormously increased as compared with the evaporation from snow-drifts and ice-fields.[7] On the other hand, in low mountain and hilly regions of humid lands, forests about the sources of the streams are of advantage in two ways: the water being in excess, increased evaporation is advantageous; and the forests serve to hold back the water and thus equalize the flow through the year and greatly mitigate the floods.

Whether forests increase the amount of rainfall has long been discussed, and lately it has been the subject of careful scientific investigation. The out-

---

[7] Powell is partly right. Dense forests, which hold large amounts of snow in their foliage, surrender it rapidly to evaporation, but so do shadeless alpine expanses. Moreover, snowpacks under forest cover yield up their water more slowly than those in open areas, producing a lower peak of runoff but a longer runoff period. Most hydrologists agree that broken, low-density forest stands produce optimum water yield: Their trees provide shade and windbreak, and their openness allows snow to accumulate on the ground and build into well-packed drifts.

come of all this research is that, if forests do, in fact, increase rainfall, it is to such a slight extent that our present means of investigation are not sufficiently refined to prove it.

Such are the facts to be collected as preliminary to the construction of a reservoir system. To neglect the essential facts is to be guilty of criminal neglect. The history of mountain-lake construction, throughout all the countries of engineering enterprise, is full of lessons like that taught at Conemaugh, and the lessons have always been enforced by the destruction of property and life; they have always been emphasized by dire disaster. Modern industries are handling the forces of nature on a stupendous scale. The coal-fields of the world are now on fire to work for man; chemical forces, as giant explosives, are used as his servants; the lightnings are harnessed and floods are tamed. Woe to the people who trust these powers to the hands of fools! Then wealth is destroyed, homes are overwhelmed, and loved ones killed.

Address to the Montana Constitutional Convention

In the summer of 1889, Powell accompanied the U.S. Senate's Special Committee on the Irrigation and Reclamation of Arid Lands on a tour of the West. On August 5, he took time from committee activities to address North Dakota's constitutional convention. Four days later in Helena, he accepted a similar invitation to speak to the framers of Montana's constitution.

Mr. President, and Gentlemen, it is with some degree of embarrassment that I speak to you after the distinguished gentlemen who have spoken today, and who have discussed national affairs so eloquently and so ably to you. In such discussion I have taken no part. I came to this country rather as a pioneer. I am older in this country than most of you. I feel as though most of you were new comers here and were tenderfeet. I came at an early time, and I want to talk now as one of the old pioneers, if you please, and not in the capacity of a statesman. (Applause.)

It has fallen under my observation from time to time and been a source—a theme of study for many years, the general question [of] what would ultimately become of this country, what would be the form of the industries in this country and the form of government in this country? For you are under peculiar conditions. In the eastern half of the United States we have settled governments, that is we have state and local governments adapted to the physical conditions of their country; but in the western half of America, the

Helena, Montana, August 9, 1889. In *Proceedings and Debates of the Constitutional Convention,* 820–23.

local, the state, the territorial, county governments and the regulations and the national government are in no sense adapted to the physical conditions of the country. (Applause.)

And these are the problems which have been interesting me for many years, and I will illustrate it by simply stating somewhat of the physical conditions of this state now forming, and explain to you what seems to me as an old pioneer, not as a statesman, if you please—what seems to me to be the form of local government and the adaptation of institutions of these physical conditions, which are necessary for your prosperity, and which you will, as I believe, ultimately have. Montana has an area of about, if my memory serves me, ninety millions of acres.[1] Of that ninety millions thirty-five millions of acres are mountainous—thirty-five millions of acres of land are dedicated to special industries thereby. In those mountains you will find silver and gold, copper, lead, and in the mountains more or less iron, in the flanks iron and coal. But the mountain region has something more in it of value, as you will see. The mountain region is the timber region of the country. On the plains and in the valleys no timber grows. The regions below are in part agricultural and in part pastural [sic]. And the remaining portion, leaving out the mountain area, about thirty-five million acres of land, can be redeemed by irrigation, according to the latest estimates and careful study of the matter, which we have made for the last two years.

It must be understood that if thirty-five million acres of land were redeemed by irrigation in this country, it means the utilization of all its waters—it means that no drop of water, which is the blood of agriculture, if you please, the life blood of agriculture—that no drop of water falling within the area of the state shall flow beyond the boundaries of the state. It means that all the waters falling within the state will be utilized upon its lands for agriculture. That is made under some careful estimate and partial estimate of the volumes of water running in your streams, a study of the rainfall of a district, and upon a further computation that it will take one acre of water to irrigate one acre of land. Now, understand what I mean by

---

[1] Today Montana reckons its area at 94,100,000 acres.

one acre of water, for as engineers we do not reckon water in inches of flow, but in acre feet. An acre of water one foot in depth we call an acre foot, and on an average in Montana for years, irrigation of an acre of land requires an acre foot of water. It is a very simple ratio as you see.

If all of the water flowing in the streams of Montana be used—not all the water flowing during the season of irrigation, but all of the water is used in irrigation, it will irrigate about thirty-five million acres; but in order to utilize all of this water and to redeem all of this land it becomes necessary to store the waters which are usually run to waste; that is, the season of irrigation in Montana will vary from six weeks to nine weeks; in some few cases it will be longer than that, but in the main we may say that the time of irrigation, the season of irrigation will be about two months. With that understanding, ten months of the flow runs to waste, and in order that the thirty-five million acres of land be redeemed for agriculture within this state, it is necessary to use all that water running to waste to the sea—to store it.[2]

Then there remains twenty million acres of land which cannot be used for agriculture, which are yet in the mountains and not covered by timber, but which yet have more or less value for pasturage purposes.

That is the condition, and, gentlemen, it is a magnificent heritage—ninety million acres of land; thirty-five million acres of mountains covered with forests, and with mountains filled with ores—thirty-five million acres of land as rich as any other that lies under the sun, to be made fertile, to be made to yield in vast abundance by the utilization of these waters. It is no misfortune as at first it may appear—it is no ultimate misfortune to the people that their land is arid. The thirty-five million acres of land which you have to redeem by irrigation will be to you much more valuable than if that water was distributed evenly over the country so that there was sufficient rainfall and irrigation was unnecessary.

---

[2] In 1997, farmers in Montana irrigated 1,994,484 acres, less than 6% of what Powell thought possible. Powell's estimate is low not just because Montanans allow water to leave their boundaries but because he consistently overestimated the West's capacity for successful irrigation. Even so, his estimates were much more realistic than those of irrigation's most passionate advocates. See note 2, p. 258, Selection 12 ("Irrigable Lands of the Arid Region").

To those who have not studied the subject with care, the proposition seems perhaps somewhat quixotic, and yet there is a long line of history to prove what I have said. I will not stop to enter upon this line of the subject, but I wish to call the attention of you gentlemen to some of the conditions which you must meet in order to secure the prosperity of these people. The question of irrigation then is undoubtedly bound up with some other questions of almost equal importance, as you will see as I proceed.

The timber of this country grows on the mountains; it is therefore not distributed throughout the country where it is needed. The agricultural people will have their timber lands five, ten, twenty, fifty and one hundred miles away from home, but the people who want to use that timber and who live down in the valleys and spread abroad over the plains, they are the people primarily interested in the forests, and the people who own the irrigable lands must own and must control the forest lands, for a part of their prosperity depends upon the utilization and preservation of those forests.

There is still another reason. In the mountainous region like those in the arid country, a large part of the rain falls upon the mountains. Usually throughout Montana you have, say, from twelve to fifteen inches of rainfall in the low country and from twenty to twenty-five, forty, fifty and even sixty inches of rainfall on the mountains. The water necessary to fertilize the agricultural lands falls somewhere else—falls upon these mountains.

And the men who are engaged in agriculture in the valleys and on the plains are interested in the mountains for another reason than that they must derive their source of timber from them. Every iota of value there is to the lands to be redeemed for agriculture depends upon the water with which they are supplied. The intrinsic value exists in the water. And the water falls not on their own land—the water falls elsewhere in the mountains, and hence the people who live by agriculture below have a double reason for being interested in the forests and mountains above, and ultimately they must control those mountain acres—the agriculturists must own and control not only the lands which they occupy themselves, but also the lands where the timbers grow, and also the lands where the waters fall which make their lands valuable.

Now, having made this brief statement of the conditions and the facts at large, let me apply them. The state is in the formative condition. You are met here to adopt a constitution, and there are questions that ultimately will be of vast interest to these people—and I beg pardon almost for mentioning them, but I have mentioned them from the standard not of a statesman but only of a pioneer—a man in the country earlier than occupants usually and you must take them only for what they are worth from an observer, and not a statesman. But this is the point. The whole area of Montana may be easily divided into drainage basin districts of country which constitute a geographical—a physical—unity; a region like the Gallatin Valley, for example, with a river flowing down through the valley with its tributaries on either side, heading out from the mountains to the very rim of the Gallatin basin on either side.

Every man who settles within the valley of Gallatin comes ultimately to be interested in every part of that valley, because it is the entire Gallatin valley, the whole drainage basin, [that] gathers the water for his farm. Only a portion of these valleys can be redeemed for agriculture; another portion will be utilized ultimately for pasturage. High up in the valleys we have the timber lands, and higher up the valleys we have the mountains where the waters are condensed. The people below must necessarily be interested in the whole drainage basin around about where these waters are gathered.

Now, agricultural industry, carried on under conditions where irrigation is absolutely necessary, is carried on under very peculiar conditions as compared with those of humid lands. You have not in this territory reached a condition of affairs where the matters which I now wish to present to you yet press upon you. They may in some few localities only; but very speedily the question of water rights—who owns this water, is to be the important question in this country. Remember the question of land rights is comparatively a minor one as compared with water rights, and water cannot be measured out to you by metes and bounds; you cannot lay out lines and drive stakes in the clouds of the heavens from whence the waters come. They pour down today in storms and tomorrow in storms, and the year in storms, and flow down to the rivers and creeks, and you have to measure them out by gallons or by acres from day to day, from month to month and from year to year. All the great values of this Territory have ultimately to be measured to you in acre feet, and in the preparation of a constitution for this great State, this great fact should be held in view.

Now, without entering too largely into the question of pointing out the necessities for regulating the use of waters and the measurement of waters, etc., I want to present to you what I believe to be ultimately the political system which you have got to adopt in this country, and which the United States will be compelled sooner or later ultimately to recognize. I think that each drainage basin in the arid land must ultimately become the practical unity of organization, and it would be wise if you could immediately adopt a county system which would be coincident with drainage basins, for in every such drainage basin you have got to have first the water courts. Disputes will arise from day to day about the waters, and in every district there must be ultimately a water court. There must be a corps of officers: water masters or supervisors who measure out this water for the people. The general government cannot, the State government will not measure the water for you, neither can they measure it for themselves, and you have got to have local self-government to manage that matter. Then, the people who are interested in these waters are also interested in the timber, and the people who are interested in the waters and agricultural lands are interested in the pasturage of those lands.

And now I come briefly to state what I believe should be done. First, I believe that the primary unity of organization in the lands should be the drainage basin which would practically have a county organization, if you please, with county courts, etc.—I need not enter into the details—then that the government of the United States should cede all of the lands of that drainage basin to the people who live in that basin. (Applause.)

I do not believe that the government of the United States can ever keep up a police force or a system of agents to manage the timber of this country.[3] (Applause.)

---

[3] At the time of this speech the principal extant model for federal timber protection was Yellowstone National Park, established in 1872, which the U.S. Army patrolled to prevent and extinguish fires, to guard against timber poaching, and, to a limited degree, to herd wandering tourists into concentrated campgrounds lest they add to the fire threat. Powell's opposition to a system of federally controlled forests was rooted in his dislike for permanent, centralized control, disconnected from the check of local self-interest, but it was not solely philosophical. An effort to reserve forest lands would compete with Powell's ongoing effort to reserve irrigation lands, especially reservoir sites.

I believe that the people of the drainage basin themselves are more interested than any other people can be in that particular drainage basin—that they are the only people who can properly administer that trust, and I believe that the people who live along every valley in this country should be the people who control three things besides the land on which they live: they should have the control of the water; they should have the control of the common or pasturage lands; and they should have the control of the timber lands (applause); and I have no doubt but that could be secured through general legislation ultimately in this country, for the interests of this western country are being rapidly understood in the east, and the people are filling up the country so rapidly that they are very soon to be able to make the eastern people respect their wants, their needs, their rights and their wishes.

The simple question which I have then to present to you, which I think would be worthy of your consideration, and which early in the history of the State or even in the adoption of your State constitution you should consider, is, what should be the primary unit of your government? And I think that the careful study of the matter will show you that the drainage basin is the natural unit and should be a county of this State. If then you will provide a system of counties by drainage basins you have the fundamental organizations, and can in time acquire all the other rights and assume all the other duties which that organization demands.

For thirty odd years I have been studying this western country when no irrigation was practical except in a few places in Utah and New Mexico and so I have been traveling about the country from year to year surveying now this portion and now that, and studying from many standpoints, and I have seen gradually agricultural industries growing up, and today within the arid region not less than seven million acres of land are cultivated by artificial irrigation, but the time will soon come when this will be multiplied tenfold and one hundredfold, for everywhere throughout the whole region west of the one hundredth meridian the people are wide awake to this problem. The mountains may be filled with gold and silver; the hills may be filled with coal and iron, but all these have little value unless there is a basis of agriculture for a prosperous people, and that you can have in all of this country. But in order that you may not have vast conflicts, in order that irrigation may be developed, that agriculture may be developed, with the least friction, with the least cost of the people, the system which I have mentioned will be of untold value.

Let me call your attention to two more points just here. The first is that litigation is a prolific source of expense and evil, and you should endeavor to provide for a proper system at the very beginning of your State for adjusting rights by your fellow citizens among yourselves. It will save you—you hardly know how much—unless you have studied irrigation in other lands you will hardly know how much, but it will save you a vast cloud, a vast amount of litigation.

Let me repeat once more to you and close with this, that you have three interests, all tied together, and they ought to be tied together in the organization of your body politic as they are tied together by physical conditions. The agricultural lands are dependent upon the mountains above, and the farmers below ought to own the control of those mountains so far as they are the source of the timber which they must use, and own and control them so far as their pasturage is concerned, for pasturage on a large scale. The range pasturage which has grown up in all of this western country[4] must necessarily decay as such and must become tied up with the agriculture of the land; not that there will be less, but more stock raising in the country then, but it will be distributed instead of in the hands of the few people—it will be distributed ultimately among all those who cultivate the soil.

I thank you gentlemen for your attention. (Applause.)

---

[4] That is, the system of range livestock ranching that has grown up throughout the West, which primarily involved large operations under the control of relatively few individuals.

oto 1. "Major Powell Inquires for the Water Pocket" by John K. Hillers, Powell Expedition, 73–75. Powell speaks with Taugu, a member of the St. George band of Paiutes. Courtesy National thropological Archives, Smithsonian Institution (photo no. 1591).

Photo 2. Major Powell (at right center) and Jacob Hamblin (to Powell's right) in council with band of Paiutes on the Kaibab Plateau, by John K. Hillers, Powell Expedition 1873–75. Courtesy National Anthropological Archives, Smithsonian Institution (photo no. 1621).

Photo 3. Major Powell (sitting against tree), painter Thomas Moran (hand to face), and compan-ions sharing a meal on the trail. By John K. Hillers, summer 1873. Courtesy National Anthropological Archives, Smithsonian Institution (photo no. 56331).

Photo 4. Major J. W. Powell in his office in the Adams Building on F Street, NW, in Washingto DC, circa 1896. Courtesy National Anthropological Archives, Smithsonian Institution (photo n 64-a-13-a).

oto 5. Geological Survey farewell dinner for Powell, 1894. From left on left side of table, sitting:
J. McGee, Marcus Baker, Grove K. Gilbert, J. C. Pilling, S. J. Kuble, Herbert M. Wilson,
W. Parker. From left of right side of table: Frank Sutton, David T. Day, C. C. Babb, Robert H.
apman, Henry Gannett, Wells M. Sawyer, Frederick H. Newell, J. Stanley Brown, John Wesley
well, William Henry Holmes, Charles B. Walcott. From left, standing: H. C. Rizer, J. S. Diller,
chard U. Goode, H. W. Turner, Robert T. Hill, William A. Croffut, DeLancey Gill, E. Willard
yes. Courtesy National Anthropological Archives, Smithsonian Institution (photo no. 43239).

PART VI

Advice for the Century

A N OLD SAYING of the rural West advises, "When you get to the end of your rope, tie a knot in it and hang on." Powell had been to the end of his rope more often than most who survive the experience. He arrived in that unhappy place in the Union hospital tents after Shiloh and in the depths of the Grand Canyon. He may have approached it on other occasions, perhaps traveling overland on the Colorado Plateau or on treks or float trips while still a teenager. Most assuredly, he reached it in the harsh days of late 1889 and 1890 after the closure of the public domain brought the Irrigation Survey and him personally under vehement attack. Powell then did more than hang on. He fought back.

The three articles included in this section represent important elements of his counteroffensive. They appeared serially in *Century Magazine* in March, April, and May of 1890—at the same time Powell was battling an antagonistic Congress to keep the Irrigation Survey intact and funded. *Century*, a lineal descendent of *Scribner's Monthly*, in which Powell had published a number of articles including those reproduced in Part III, was a widely read general interest magazine, much like *Harper's* or *The Atlantic Monthly* is today. It provided an ideal vehicle by which Powell might communicate directly to the American public the essence of what he was testifying, often under hostile cross-examination, in committee rooms on Capitol Hill.

But the importance of the *Century* articles goes beyond the immediate debate of which they were a part. The articles, which never specifically mention the Irrigation Survey or its predicament, do not just take aim at the needs that brought the survey into being and necessitated its continued existence. They also address the needs that would remain, should the survey continue, even after its work was accomplished. They call for water courts and interstate adjudications; they call even for a wholesale reformation of political organization in the West.

The *Century* articles are a kind of sequel to *The Report on the Lands of the Arid Region*. In them one finds the fullest development of Powell's original ideas, now matured another dozen years. There are differences in detail—on the sources and impact of wildfire, for instance; and there are differences in underlying substance—notably in the institutions Powell recommends, about which more in a moment. But the two efforts are of a piece. The enduring significance of the *Arid Lands* report derives from its synthesis of diverse analyses of climate, geography, land use, land law, and other subjects into a coherent vision of how things were in the West and how they should be made to be. In the *Century* articles that vision becomes even sharper and more penetrating.

Unfortunately, Powell does not always get quickly to the point, and we may wonder at some of his literary devices—the stiff, pseudo-Indian verses, for instance, that commence the article on nonirrigable lands. Powell is keenly aware that "numbers perhaps are more arid than land," and so he tries, at times too hard, to engage the reader. But we must also appreciate his challenge. At a time when the word *ecology* had scarcely entered the language, he endeavored to educate the public about complex ecological relationships—between, say, forests, grass fuels, grazing, and wildfire. American society was still in the earliest stages of developing what might be called an environmental point of view, and few of Powell's potential readers were prepared to appreciate, let alone understand, the kinds of issues that drove his analyses and recommendations.

This is not to say that Powell's understanding of these issues was unflawed. His readiness to cut down trees to increase water yield alarmed conservationists of his day, who in many cases portrayed Powell as the enemy of *all* forests, which he demonstrably was not. He was, however, resolutely utilitarian in his approach to natural resources. Where forests were concerned, he urged the protection of stands of merchantable timber needed for human use. He also advocated protecting forested buffer zones to prolong the life and ensure the water quality of the myriad reservoirs he hoped to see built. But, ignoring other values, he viewed forests that fit neither category as somewhat expendable, for he believed that open lands would yield more of the environmental variable he most wanted to maximize: water. Such a position angered members of America's nascent forest protection movement, who conversely believed that only unimpaired, "fully stocked" forests could produce the desired abundance of water, as well as the more humid climatic conditions that would maximize water supplies.

It was an argument that forest protectionists like Bernard Fernow and Gifford Pinchot eventually won, but not because they were right about the conditions for producing optimal water supply. Their school of thought, transmitted through time, gave birth to numerous schools of forestry, including Yale's, which opened its doors in 1900, and to notions of forest ecology, now discredited, that were taught in those schools until recently. They also helped bring into being the United States Forest Service and Smokey Bear. The central message common to these venues and entities was that trees were good, more trees were better, and any land that could grow trees should be made to do so. Of course, we know better now—or at least we think we do. We know that fire, contrary to Smokey's adjuration, is a useful tool of forest management and an essential element in many ecosystems; we know that too many trees can diminish streamflow and water yield as well as endanger forest health; we know that in certain instances grasslands resist erosion better than the forests or woodlands that would replace them; we know that grasslands and open spaces, even in forested zones, are not lapsed forests, as Powell's adversaries depicted them, but part of a natural and dynamic landscape mosaic. All these understandings are present, if embryonic, in Powell's ecology and largely absent in that of his adversaries.

The Major may have ground his axe too long and hard about the unsacred-ness of trees, but he was right to keep the blade sharp.

If forests were not particularly sacrosanct in Powell's hard-eyed view of nature, neither were rivers and streams. Indeed, the toughest obstacle in making Powell an environmental hero is to place his fervor for building dams and reservoirs in some kind of nature-friendly context. Wallace Steg-ner argued that Powell would never have approved the growth of the Bureau of Reclamation into a dam-building juggernaut and that his ten-dency to take the long view would have led him to conclude that "consider-ations such as recreation, wildlife protection, [and] preservation for the future of untouched wilderness, might sometimes outweigh possible irriga-tion and power benefits."[1] Stegner was probably right that Powell's demo-cratic idealism would have led him to condemn Reclamation's early aban-donment of social goals and its metamorphosis into an agent of giant industrial projects. Nevertheless, one looks in vain for evidence that he would have valued a live river much beyond its instrumental worth as a con-duit for irrigation water.

Other friends of Powell assert that his insistence on local control and financing would have guaranteed his opposition to giant dams like Hoover and Glen Canyon, which impound Lakes Mead and Powell, respectively, on the Colorado River. Such colossi, they say, could never have been built without the active financial participation of the federal government, and Powell would not have tolerated that. Maybe so, and yet in the first article of the *Century* trio, Powell clearly contemplates "great works . . . costing mil-lions of dollars" on the lower Colorado. Another argument made on Pow-ell's behalf is that he favored the construction of many small reservoirs in the highlands over giant, wastefully evaporative lakes lower down. This is emphatically true, and on this basis it is hard to believe that Powell would hold much love for Glen Canyon Dam or the vast reservoir behind it (known to some as Lake Foul) that is named for him.

The problem of small reservoirs nevertheless remains. To say Powell

---

[1] Stegner, *Beyond the Hundredth Meridian,* 361.

wanted them to be numerous understates the case. Powell was beaverlike in his urge to impound flowing water. In the Valles Caldera of New Mexico, a spectacular mountain bowl filled with grassy valleys, Powell wanted to build not one reservoir but two, essentially checking the natural outflow of the basin.[2] In July 2000, to the delight of environmentalists who wanted to see the wildlands of the caldera protected from subdivision and development, the federal government and a nearby Indian pueblo bought the property from private owners for the handsome price of $101 million.[3] Probably neither the enviros' delight nor the buyers' motivation would have been as great if Powell's two reservoirs had shimmered there. And if Powell had had his way throughout the West, a similar diminishment of delight would have obtained in hundreds of other flooded mountain valleys, as well as downstream in the dry beds of their creeks and rivers.

This is not to say that Powell failed to appreciate the beauty and grandeur of free-flowing rivers. Page after page of his *Exploration of the Colorado River* is imbued with awe, wonder, and admiration for the majesty of the wild Colorado. To understand him and his detached, unsentimental view of western lands, we must remember that in his day almost no western rivers had been dammed, and few if any forests had been placed under systematic management. The continued settlement of the region would necessitate both measures. It is impossible to say how far Powell would have gone in harnessing rivers and manipulating forests before being checked by a desire to preserve their natural integrity. The question itself is anachronistic.

A more important question in appraising Powell is whether, in order to appreciate him in his own time, we must make him a hero in terms of our own. Powell was not an environmental saint, and no one need try to make him one. But he was utterly remarkable in his day for his insistence on understanding the relationship between the institutions of human society

---

[2] "Ceding the Arid Lands to the States and Territories," House Report 3767, 51st Cong. 2nd Sess. (1891), Serial No. 2888, 35–36.
[3] Santa Clara Pueblo, using private funds, purchased 5,045 acres of the 94,000-acre Baca Ranch, which comprised nearly all of the caldera. The federal government purchased the rest.

and the environment in which that society dwells. For him, the lands and waters of the Arid Region were no mere stage on which society's actors might freely play out their roles. He saw the physical environment as a force, unforgiving and powerful, that would shape society whether society wanted the shaping or not. His advice, repeated to his followers and critics with equal courage and conviction, was that society should adapt itself to the imperatives of that force to preserve the benefits of the land and then share those benefits equitably among its people. One wishes such a goal might have more widely pervaded Powell's time. Indeed, one wishes it would infect our own.

Powell tended to see the connectedness of things in social as well as ecological terms. The most revolutionary of all his conceptualizations is his idea of watershed commonwealths, of which he envisaged the formation of as many as 150,[4] and it is here that we encounter the greatest difference between the *Century* articles and the *Arid Lands* report. The earlier report called for the formation of numerous grazing cooperatives structured similarly to irrigation districts, and it would have placed timberlands under the control of "lumbermen and woodmen." By 1890 Powell's view of how best to govern western lands and resources had changed. In the third of the *Century* articles, "Institutions for the Arid Lands," he states clearly that "the plan is to establish local self-government by hydrographic basins." He had enlarged his unit of organization from local cooperative associations to entire watersheds, perhaps because he realized that he had earlier asked for more in the way of voluntary self-organization than the public was willing or able to give, and perhaps also because he realized the value of creating broader, diverse management units whose interlocking interests would act as checks against resource abuse.

In the 1890 articles Powell's understanding of the interrelationship of lands within the political unit has also advanced. He calls for the manage-

---

[4] "Ceding the Arid Lands to the States and Territories," 88–89. Although Powell uses an estimate of 150 in his testimony and elsewhere, the number of shaded watershed subdivisions in the map "Arid Region of the United States, Showing Drainage Districts" is closer to 140. See map 7, this volume.

ment of all three major land classes (as he defined them: irrigable, pasture, and timber lands) to be integrated under the control of each commonwealth. Wisdom would rule the use of those lands—or at least have a chance to rule it—because decisions would emanate from "a body of interdependent and unified interests and values." The commonwealths, as Powell explained in his address to the Montana Constitutional Convention (see Selection 11), would be roughly equivalent in political heft to counties, which they would replace. They would administer grazing and timber lands as a kind of commons, held for the benefit of all within the commonwealth. Interestingly, however, Powell would not have the commonwealths receive title to these lands, lest they sell or otherwise dispose of their commons and break the unity of relationships binding the lands in a whole. Instead, the United States would retain sovereign ownership of the commons of the Arid Lands and would serve as their trustee, holding the lands in perpetuity for the benefit of the commonwealths.

This is a remarkable vision and well worth considering today in the early years of the twenty-first century. Institutions for managing public lands in the American West continue to evolve, albeit slowly. In most of the region, the level and intensity of public involvement in the development of management goals, if not decisions, continue to increase. Although no two formulations of the latest managerial grail, ecosystem management, seem to be exactly the same, most of them trend in the same direction: toward recognition that no long-term management regime, however technically correct or grounded in the "best" science, is likely to succeed without public support and understanding. Accordingly, nearly all such undertakings claim to seek a high level of public involvement and a strong sense of local ownership, watershed by watershed.[5] It is hard to avoid the impression that matters are headed, albeit falteringly, in the direction that Powell urged more than a century ago. Much of the western landscape is still held in trust by the federal government. Watershed polities have rarely been formed, but in recent years watershed-

---

[5] See Johnson et al., *Ecological Stewardship,* specifically deBuys, Crespi, Lees, Meridith, and Strong, "Cultural and Social Diversity and Resource Use," III: 189–208.

based interest groups, some with a modicum of government sanction and varying levels of clout, have become an increasingly common feature of the political landscape.[6] They have come into being for the simple reason that Powell said they must: because problems touching society's relationship to its environment do not obey the limits of arbitrary political boundaries. When those problems are contained, they tend to be contained or confined bioregionally, and watersheds are the simplest of all bioregional units.

Because we have moved so far in the direction of Powell's recommendations and because we are still moving that way, it is worth asking what would have happened if Powell's plan had been accepted and implemented when it was first proposed.

First, Powell's plan would not and could not have remained unchanged. The watershed commonwealths would have developed—one hopes adaptively—in parallel with the burgeoning settlement within them. Powell certainly did not anticipate the degree of urbanization and industrialization that has transformed the West since his day. Who did?

Second, nothing in Powell's plan suggests specific treatment or protection for the land interests of Indians, Hispanics, or other minorities. If he contemplated special provisions to defend tribes and land grant heirs from dispossession, he does not mention them, so we can probably assume that his plan would not have greatly altered their treatment by the rest of society, except in one important regard. Within the framework of watershed commonwealths and through the exercise of majority rule, populous indigenous groups might have had a fighting chance to preserve their traditions of communal ownership and use of land. In the land grant country of northern New Mexico, for instance, the privatization of land grant commons—one of the severest historical injuries suffered by the people of the region—might have largely been prevented.

Third, the watershed commonwealths would not have been environmental Shangri-Las. Land abuse might not have proceeded as rampantly as it did in the open-range days of the public domain, but the commonwealths

---

[6]See the Introduction to Part IV, note 4, p. 145.

nevertheless would have experienced a share of it. Powell assumed that the desire of farming communities for abundant and high-quality water would drive decisions on use of pasturage and timberlands, but almost certainly he overestimated the importance of agriculture and underestimated the influence of towns, manufacturing industries, sawmills, and mines, all of which might have looked greedily at the resources of the commons, especially timber. Powell, like most of his contemporaries, never fully appreciated the burgeoning power of corporations, and sooner or later his watershed democracies would have needed protection from that quarter. From another vantage, environmentalists might further point out that there is no room in Powell's plan for such a concept as protected wilderness. Then again, in Powell's day there was little room in anyone's plan for such an idea, the very concept of which had scarcely been enunciated. The closest equivalent lay in the creation of national parks, a program still in its infancy. As of the dawn of 1890, only Yellowstone existed; more were added that year, and Powell himself advocated the protection of the lands that eventually became Grand Canyon National Park.[7]

But this is not to say that the watershed commonwealths would have failed environmentally. If Powell had had his way, there might have been 150 of them—150 separate experiments in land management, with 150 different outcomes. Many would have produced unfortunate results, not unlike those that actually befell almost every region of the West. Others might have distinguished themselves with foresight, wisdom, and a gentle touch. We will never know. All we can be sure of is that the variety of results would have been greater than that which the Arid Lands have thus far known. And this, in turn, would have generated, albeit with no guarantee of realization, the promise of faster and better learning about the true character of western land and society's relationship to it.

A fourth consequence of the implementation of Powell's plan would have been an expansion of the meaning and experience of democracy. Imagine 150 separate elected local governments hammering out policy for

---

[7]Worster, *A River Running West*, 471.

the use of common lands. Imagine the long debates and longer community meetings. Imagine the huge investment of civic time and energy required to manage so much land of such importance to so many people. Thomas Jefferson would have grinned at the prospect. Again, the outcome must be regarded as unpredictable, not least because there would have been so many outcomes. And that is what is tantalizing in this game of What-If: A profusion of experiments in local resource management democracy, at the scale Powell contemplated, would have generated a range of fascinating results.

Finally, if Powell's plan had been implemented, the tenor of life in the West would be different. Among other things, westerners would have had to find something else to complain about besides the federal government. That in itself might have unleashed a welcome flood of otherwise untapped human creativity. Powell's position was clear. He believed that once the central government had established fair rules for economic and social interaction, it should get out of the way. People would have to be responsible for themselves: "Furnish the people with institutions of justice, and let them do the work for themselves." His stance was parental but not paternalistic: "With wisdom you may prosper, but with folly you must fail."

Who would have prospered? Who would have failed? The outcomes lie an unknown distance down a road not taken. Short of getting a second chance at settling the West, we will never know. But as long as people read the work of John Wesley Powell and reflect on his wisdom, we will never stop wondering. And as long as westerners struggle to understand not just what might have been but what should be, we can never stop imagining.

The Irrigable Lands of the Arid Region

Powell and the Irrigation Survey were under attack in Congress and elsewhere. Powell responded by presenting his ideas directly to the American people through a series of articles in one of the nation's most widely read magazines. Here the series begins.

Nearly half of the lands of the United States, exclusive of Alaska, are arid. By this characterization it is meant that the rainfall is insufficient to fertilize crops from year to year. In favorable seasons some of these lands receive sufficient rains during the months of growing vegetation to produce fair crops, but the years are infrequent when such conditions prevail, and the areas thus favored are not of great extent. That arid lands may be available to agriculture it is necessary that they be artificially supplied with water; and this is called irrigation. Every farm, orchard, vineyard, and garden is dependent upon an artificial supply of water. The tree on the lawn, the rose on the parterre, and the violet on the baby's grave must have some loving hand to feed it with the water of heaven or it withers and dies.

When the farmer sows his field and waits for the rains of heaven to fertilize it, if the clouds are kind and come with gentle showers, he reaps a bountiful harvest; but when the heavens are as brass, famine stalks abroad, and when storms desolate the land, he plants in vain. But in the western half of the United States physical conditions like those of ancient Egypt and Assyria prevail. The clouds no longer fructify the fields with their showers.

They rarely hover over the valleys and plains where the fields and gardens lie, but they gather about the mountains and hurl their storms against the rocks and feed the rivers. The dweller in the valley waits not for showers, or waits in vain; for the service of his field's rivers must be controlled.

But will not the hills of New England, the mountains and plains of the sunny South, and the prairies of the middle region be sufficient for the agricultural industries of the United States? The area is vast, the soil is bountiful, and the heavens kindly give their rains; why should the naked plains and the desert valleys of the far West be redeemed? Why should our civilization enter into a contest with nature to subdue the rivers of the West when the clouds of the East are ready servants?

Gold is found in the gravels of the West; silver abounds in the cliffs; copper is found in the mountains; iron, coal, petroleum, and gas are supplied by nature. The mountains and plateaus are covered with stately forests; the climate is salubrious and wonderfully alluring. So the tide of migration rolls westward and the arid region is being carved into States. The people are building cities and towns, erecting factories, and constructing railroads, and great industries of many kinds are already developed. The merchant and his clerk, the banker and his bookkeeper, the superintendent and his operative, the conductor and his brakeman, must be fed; and the men of the West are too enterprising and too industrious to beg bread from the farms of the East. Already they have redeemed more than six million acres of this land; already they are engaged in warfare with the rivers, and have won the first battles. An army of men is enlisted and trained, and they march on a campaign—not for blood, but for bounty; not for plunder, but for prosperity.

But arid lands are not lands of famine, and the sunny sky is not a firmament of devastation. Conquered rivers are better servants than wild clouds. The valleys and plains of the far West have all the elements of fertility that soil can have. As the blood in the body is the stream which supplies the elements of its growth, so the water in the plant is its source of increase. As the body must have more than blood, so the plant must have more than water for its vigorous growth. These conditions of plant growth are light and heat. While the roots of the plant are properly supplied with water and other elements of plant growth, the leaves must be supplied with air and sunshine. The light of a cloudless sky is more invigorating to plants than the gloom of storm. Abundant water and abundant sunshine are the chief conditions for vigorous plant growth, and that agriculture is the most successful which best secures these twin primal conditions; and they are obtained in the highest degree in lands watered by streams and domed by clear skies. For these reasons arid lands are more productive under high cultivation than humid lands. The wheatfields of the desert, the cornfields, the vineyards, the orchards, and the gardens of the far West, far surpass those of the East in luxuriance and productiveness. In the East the field may pine for delayed rains and the green of prosperity fade into sickly saffron, or the vegetation may be beaten down by storms and be drowned by floods; while in the more favored lands of the arid region there is a constant and perfect supply of water by the hand of man, and a constant and perfect supply of sunshine by the economy of nature. The arid lands of the West, last to be redeemed by methods first discovered in civilization, are the best agricultural lands of the continent. Not only must these lands be redeemed because of the wants of the population of that country, they must be redeemed because they are our best lands. All this is demonstrated by the history of the far West, and is abundantly proved by the history of civilized agriculture. All of the nations of Egypt were fed by the bounty of one river. In the arid region of the United States are four great rivers like the Nile,[1] and scores of lesser rivers, thousands of creeks, and millions of springs and artesian fountains, and all are to be utilized in the near future for the hosts of men who are repairing to those sunny lands.

There are nearly 1,000,000,000 acres of these arid lands in the United States, of which nearly 120,000,000 acres can be irrigated when all such waters are used. Already more than 6,000,000 acres are cultivated through the agency of canals. Thus the experiment has been tried, and doubt no

---

[1] Powell refers to the Colorado, the Snake and Columbia system, the Missouri, and the Rio Grande.

longer rests upon the practicability of western irrigation. It is fully demonstrated that the redemption of these lands is profitable to capital and labor. An acre of western land, practically worthless without irrigation, when the works are constructed to supply it with water at once acquires a value marvelous to the men of the East. In new California, settled but yesterday, cultivated lands command better prices than in Massachusetts or Maryland, and this is because an acre of land there will produce two or three fold the quantity of food for man or beast that an acre will here, for the average year. We of the East must recognize that while the lands of the West are limited in quantity to comparatively small and level tracts in the valleys and plains which can be served with water by canals, yet the limit in quantity has compensation in quality.

To accomplish the redemption of the arid region capital in large amounts is needed. Some lands can be reclaimed at a cost of two or three dollars an acre; others, ten or twelve; while in some cases, where lands are of great value by reason of their proximity to cities, hundreds of dollars per acre will be expended to bring waters from distant springs or from the depths of the earth. A rough estimate may be made that 100,000,000 acres can be redeemed at the rate of ten dollars per acre—that is, for 1,000,000,000 dollars. In this work vast engineering enterprises must be undertaken. To take the water from the streams and pour them upon the lands, diverting-dams must be constructed and canals dug.[2]

---

[2] In 1997, according to the U.S. Department of Agriculture (www.nass.usda.gov/census/census97), the total irrigated area of the 17 western states (excluding Alaska and Hawaii) stood at 42,961,483 acres. This amount includes more than 16 million acres in states that straddle the hundredth meridian: Kansas, Nebraska, Oklahoma, Texas, and the Dakotas. A fair portion of irrigated lands in those states lie east of the line that Powell used to define the Arid Region, and many of their irrigated lands are watered by groundwater pumping, a means of irrigation that did not figure heavily in Powell's calculations. Although exact figures are not readily available, the actual extent of irrigation agriculture in the Arid Region surely does not today exceed 40 million acres, far less than the 100–120 million acres Powell predicted and the vaster area on which the impatient and intolerant delegates to the 1893 Irrigation Congress placed their hopes. Powell's estimate of the cost of irrigation development is still more difficult to evaluate in contemporary terms, given the ever-changing value of the dollar, but it is certainly lower than the amount that has actually been expended. In the

With most streams the water is insufficient to serve the lands, and a selection must be made. The conditions which should govern this selection, though somewhat complex, are of grave importance. The rains fall chiefly on the mountains and high plateaus, where the lands are nearly or quite valueless for agriculture. Cliffs, gorges, and steep declivities are not attractive features to the farmer. At great elevations snows fall and accumulate in vast fields, deep drifts, and icy glaciers, and linger long through the spring, sometimes remaining all summer. On these elevated lands late June and early September frosts come, and the days of July and August are not wholly exempt from their ravages. Thus the elevated lands are not attractive to agriculture. The farms, hamlets, towns, and cities have their sites away below on the sunny lands. Here and there mines of gold and silver attract a population and induce men to build homes in the upper region of snow. But their supply of food must come mainly from below. The mountain streams while yet small, as brooks and creeks, cannot be used to advantage, and when they leave the mountains they are in most cases already great creeks or rivers. A mountain stream flows in a deep, narrow gorge, down which torrents of water roll in mad energy. Such is the crystal river of the mountains. When it strikes the plain it is suddenly transformed. The steep declivity is changed to one of low degree, and a deep, narrow stream spreads into a broad sheet of water ten, twenty, fifty times as wide as above. When the river is thus transformed it undergoes another change; on the plains below it gathers the sands and dust, and the deep, crystal stream becomes a shallow river of mud. Such are the characteristics of the greater number of streams of all the arid region.

The place of transformation, where the mountain stream of pure water is degraded into a lowland stream of mud, is an important point when the

---

[2](continued) 1950s, for instance, senator Paul Douglas of Illinois, a critic of western water projects, concluded that the per-acre cost of reclamation varied, project by project, from $674 to nearly $4,000—a far cry from Powell's estimate of $10, even allowing for inflation. (See Reisner, *Cadillac Desert,* 148.) More recently, the Central Arizona Project, which carries Colorado River water across the tablelands of northern Arizona to the metropolises of Phoenix and Tucson, carried a price tag, by itself, of $2 billion.

stream is to be used in irrigation.[3] If the waters are turned out in the valleys above, they are used where they will perform the least service, for the climate is unfavorable to agriculture. Such lands are chiefly valuable as pasturage. Grass, potatoes, and rye, and in general the crops of Norway and southern Alaska, may be cultivated with some success; but, in sight of the sunny plains below, it is a waste of water to use the rivers in these regions of ice. On the other hand, the streams cannot be used with the greatest advantage far down their course and distant from the mountains. The storm-waters and fierce winds of the low plains and valleys, that are arid and dusty for most days of the year, fill the valleys and shallow channels of the mud-bearing rivers with vast accumulations of sand. In these broad stretches the waters spread and are largely lost by evaporation. Very many of the streams of the arid regions, perhaps two out of three, are thus swallowed up by the sands, and are called "lost" rivers or creeks. Others have a sufficient supply from the mountains during seasons of flood to enable them to cross the hungry sands, and deliver a part of their volume to lower channels in more humid lands, through which they find their way to the sea. They die in seasons of drought and live in seasons of storm. Still other rivers flow perennially but dwindle on their course over the dry plains. The "lost" streams must be used near to the mountains or not at all. The intermittent streams and the diminishing rivers should be used near to the mountains before a large part of their waters is lost. A stream that will irrigate a million acres of land near the mountains would be sufficient to serve only two or three hundred thousand acres a hundred miles away. There are other reasons why the river should be taken out from its channel where it emerges from the mountains.[4] At that point diverting-dams can be constructed with the least expense and maintained at the least cost, and be made to command lands to the greatest advantage in the construction of minor canals; while the waters

---

[3] Powell's eastern readers would have had little understanding of western landscapes (for instance, how emphatically climate varied with elevation or how rivers changed character as they flowed from mountain to plain). In the passage that follows, Powell provides an introduction to the physical geography of the region.

[4] Implicit here is the idea that virtually all of a river's water should be put to agricultural use.

below, when charged with great quantities of sand, speedily destroy the works of irrigation, and the sands injure the fields.

Most irrigated lands ultimately require drainage. The bottom lands of the great rivers soon become filled with water, and are transformed into swamps and destroyed for the best agriculture. The low plateaus are ultimately far superior to them for all agricultural purposes. Thus it is that the higher lands away from the rivers and near to the mountains should be first served. Only a part of the water poured upon lands for their irrigation is evaporated to the heavens; another, and perhaps larger, part returns to the river. The irrigation of the upland creates many springs, which unite to form brooks and creeks, and the waters can thus be used again and again, but in diminishing quantities.[5] A proper system of drainage not only improves the land drained, but conserves the water to be used again. It is thus that with every system of supply-canals a related system of drainage-channels and canals must be planned for the benefit of the fields first irrigated and for the increase of the area of irrigation.

The season of irrigation is short, varying in different latitudes and altitudes from two to five months. In some regions of country the season of flood precedes and extends into the first part of the season of irrigation; in other regions floodtime comes late, when the time for supplying water is nearly past. In a few cases maximum supply and maximum want are coincident in time. In all cases where they are not synchronous the excess runs to waste; the unused waters are lost in the sea. During all the months when irrigation is not in progress the entire volume is unused, if the only struc-

---

[5] The concept Powell introduces here is today known as consumptive use: the idea that evaporation and vegetative transpiration consume only a portion of the total amount of water applied to irrigation use. The unconsumed portion constitutes return flow and remains available for downstream recapture and reuse. One problem Powell does not mention is that with repeated use the amount of dissolved minerals in water (i.e., its salinity) steadily increases, with the result that the water's suitability for agriculture and other uses can be compromised. At various times in the 1960s, for instance, Colorado River water delivered by the United States to Mexico was so saline it killed crops rather than nourishing them. Under a 1973 amendment to the 1944 water treaty between the two nations, the United States must today ensure that the water it delivers across the border meets standards for freshness.

tures are diverting-dams and canals. To save this water reservoirs are needed. In their construction and the selection of their sites many interesting problems are involved. Some of the conditions which govern the selection of sites are of great importance. Evaporation from the surface of water varies, under different climatic conditions, from thirty to one hundred inches. A reservoir most favorably located may lose less than three feet of water during the year, while, under most unfavorable conditions, the loss may be more than eight feet annually.[6] Evaporation is greater in the hot, dry lands below and less in the cold, humid lands above. The law of diminution is complex, having many factors, and is not yet very well known, but the general statement made is substantially correct. For this reason storage-reservoirs should be constructed in the mountains. In many of the northern ranges of the West favorable sites are found. Already many mountain lakes exist that can be used for this purpose by deepening their outlets and constructing gateways, so as to permit the lakes to be filled when the waters are not needed and to be tapped when a supply is demanded. There are many mountain valleys that are morainal basins admirably adapted to this purpose, and where reservoirs can be constructed at small cost. The mountain regions of the West have many lakes of cold, emerald waters, and these are to be multiplied by the art of man and made to hold the waters needed to refresh the arid plains below—treasure-houses where the clouds are stored.

The mountain ranges of the western portion of the United States differ very greatly in their topographic characteristics. Sometimes advantageous reservoir sites can be found in the upper regions; sometimes low valleys, or parks, are found nearly inclosed by mountains and foothills, while there are many ranges which have such steep declivities and terminate so abruptly on the plains that sites are infrequent. For such reasons not all of the mountain waters can be stored in mountain lakes, and it becomes necessary to construct reservoirs on the plains below. Here the streambeds cannot be uti-

---

[6] Actual development of large-scale water projects in the West has largely ignored Powell's caution about evaporation. For example, evaporative losses from reservoirs on the main stem of the Colorado River, all of them necessarily located at low, hot elevations, exceeds 1.5 million acre-feet per year, which is roughly equivalent to one-tenth of the river's annual flow.

lized, because of the difficulty of maintaining works on broad flood-plain lands composed of incoherent sands, and because the muddy waters below discharge their silt and fill the reservoirs with great rapidity, so that the life of such a reservoir is too short to warrant the expense of its building. Under such circumstances a river should be turned from its natural course into a canal near the point of transformation, and be conducted into some lateral valley which has been excavated by storm-waters. In general, favorable sites of this character are frequent. The valley is utilized by selecting some point where the inclosing hills converge, and there constructing a retaining-dam.

When all the perennial waters of springs, brooks, creeks, and rivers are used by canals and reservoirs, the total supply of available water for irrigation is not exhausted. All of the arid lands below have some rainfall, varying from three to twenty inches, from year to year, and from region to region. The rains which fall upon these thirsty lands are in part absorbed and ultimately evaporated, but often the storms come with great violence, and local floods arise therefrom. These storm-waters can be caught and stored in basins among the hills and used for agricultural purposes. The amount of water that can thus be saved is no mean quantity. But it must often be stored in small reservoirs of a few acres each; and this means the construction of ponds on farms, scattered here and there among the hills where sites are favorable; and the waters will thus be used on small tracts of land distributed far and wide over the arid plains and valleys. Ultimately the whole region will be covered with a mosaic of ponds fringed with a rich vegetation; and crystal waters, and green fields, and blooming gardens will be dotted over all the burning, naked lands, and sand dunes, alkali stretches, and naked hills will be decked with beautiful tracts of verdure.[7] Not all the storm-waters will thus be caught; much will still fall into the great sand valleys and flood-plains, and there disappear in the sands; but such valleys have a floor of solid rock; and so the waters are stored in the silt of ancient

---

[7] Hundreds of thousands of stock tanks scattered across the rangelands of the West, most created by construction of an earthen berm across a swale or arroyo, embody the fulfillment of Powell's advice. "Beautiful tracts of verdure," however, are rarely associated with them, either by reason of grazing pressure or limitations in the site.

floods, where they may be brought to the surface again by pumps and other hydraulic devices, and be made to irrigate many a stretch of farm land.

There is one more source of water. In the flexing of the strata of the earth through geologic agencies subterranean basins are formed, where rocks below, impervious to water, are separated by water-bearing strata from the rocks above through which the water will not pass. Into these water-bearing strata wells may be sunk, and the water will often flow to the surface. Such artesian wells are often used in irrigation, and they will be used to a much larger extent in the future. Artesian waters are not found everywhere in the country, but only in geologic basins, and to select sites for them a knowledge of the geologic structure is necessary.

By the use of all the perennial streams during the season of irrigation, by the storage of the surplus water that runs to waste in seasons when irrigation is not practiced, by the impounding of the storm-waters, by the recovery of the floods accumulated in valley sands, and by the utilization of the artesian fountains, a vast area of the arid lands will ultimately be reclaimed, and millions of men, women, and children will find happy rural homes in the sunny lands.

From the brief account given it will be seen that in order to redeem the arid lands it becomes necessary: first, to select properly the lands to be redeemed; secondly, to select the reservoir sites where the water is to be stored; thirdly, to select canal sites, and these should be dedicated to public use, so that individuals may not acquire title to the lands for the purpose of selling them to the farmers when the irrigating works are to be constructed, and thus entailing upon agriculture an unnecessary expense;[8] fourthly, the extent and method of utilizing the flood-waters stored in the sands must be determined; fifthly, the artesian basins must be discovered and their extent and value revealed.

---

[8] Powell is referring to the kind of speculation that prompted the Department of Interior to close the public domain to new entry in August 1889, slightly more than half a year before this article was published. The ensuing storm of controversy imperiled (and ultimately doomed) Powell's irrigation survey, which he here proceeds to justify.

For this purpose there are necessary:

(a) A topographic survey, that the mountains, hills, and valleys may be outlined and their relative levels determined, and the whole represented on appropriate maps.

(b) A hydrographic survey. The waters of the streams must be gauged, in order to determine the volume which they carry through the different seasons of the year. Then the rainfall must be determined, for the amount of water to be supplied by canals is supplementary to this. Where the rainfall is twenty inches a small artificial supply serves the land; if it be but five inches a large supply is necessary. Then the amount of precipitation for various sites of reservoirs must be determined, to discover the amount which can be saved. And finally, it becomes necessary to determine the amount of water which is needed to serve an acre of land. This is called the "duty" of water, and in the United States it varies widely. In some regions of country, where the rainfall is great and the soil favorable, the duty of water is large: a given amount of water will irrigate a broad tract of land. But where aridity is excessive and the soils are unfavorable, such given amount of water will irrigate but a small tract. For the purpose of measuring stored water many engineers have come to use an "acre foot" as a unit, which means an acre of water one foot in depth. In some portions of the United States an acre foot of water will irrigate two or three acres of land for one season; in other regions two acre feet are necessary to the acre; but these are extreme conditions. The general average, which largely prevails, may be stated as an acre foot of water to an acre of land;[9] and a lake which contains 100,000 acre feet of water will serve 100,000 acres of land for one year. In the practice of irrigation it is found that it takes two or more years properly to fill the ground with water, and for these first years a much larger supply than has been indi-

---

[9] Powell is far from the mark here. Farmlands in the Imperial Valley of California, one of the hottest irrigated regions on the continent, use between 5 and 6 feet of Colorado River water a year. Diversions for cooler lands generally are less but rarely so little as 1 acre-foot per acre. Actual consumptive use may approach 1 foot per acre, but given Powell's reference in this passage to upstream storage, he seems to be discussing the amount of water to be diverted, not merely that which is consumed.

cated is necessary. Where a supply has been secured for 100,000 acres by reservoir or canal, the lands which it will ultimately serve can be redeemed only through a course of years. Perhaps a third or a half of the land can be supplied for the first year, and to this new areas can be added, from season to season, until at last the whole duty of water is secured.

(c) An engineering survey. The reservoirs, canals, and ancillary appliances must be planned and their cost estimated.

(d) Finally, a geologic survey, to utilize the waters of the sand reservoirs and artesian wells.

Such are the scientific problems involved in the redemption of the arid lands.

A brief survey of some of the more important irrigable districts of the West will serve to set forth other interesting facts relating to this subject. In central Colorado the "Continental Divide" is a wilderness of desolate peaks that rise far above the timber line into regions of rime and naked rock. Here, with other rivers, springs the Arkansas, in deep cañons and narrow rocky valleys. Many silver creeks, with water flashing in cascades, unite to form a river which plunges down a steep mountain valley until it passes the foothills and spreads in a broad, turbid stream at the head of the great valley of the Arkansas. Then it creeps over the sands in tawny ripples, down the incline of the plains, becoming less in volume by evaporation and the absorption of the waters in the sands, but growing in size from the accession of smaller tributaries that come from distant mountains on either hand. After crossing the Colorado line it grows perceptibly smaller until a more humid region is reached, where other tributaries join it, and it soon becomes a great river. In the stretch that begins just above the State line and extends across Kansas its channel often becomes dry, and the sands drift in the winds from bank to bank. But in seasons of flood a broad, shallow torrent rolls across Kansas into the State of Arkansas and bears along to the lower region vast loads of mud, choking the navigable stretch below with "sand-bars," that act as dams, by which the floods are turned over the valley, and the fields are ofttimes destroyed. Already the farmers of Colorado have taken the water on their lands, and the river is made to do duty to its utmost capacity in seasons of drought. But the surplus waters yet run to the sea. Some of them can be stored on the plains; but the land available for irrigation is far in excess of the amount which the river can serve. Where shall this water be used? If in the mountain valleys, it will largely be wasted; if in the great valley below, how shall it be divided between Colorado and Kansas? It is worth millions of dollars annually. To whom shall it be granted? If the larger part is to be used in Colorado, how shall it be divided between the several districts through which it passes? The law is practically silent on the subject. Heretofore every man might help himself; but at last the question has arisen, controversies have sprung up, and the States are almost at war.[10]

The Rio Grande flows through San Luis Park,[11] where there is a great body of comparatively level land. Here the waters have been taken out and many hundred thousand acres irrigated. Neglecting the tributaries, let us follow the river across the line into New Mexico. Again the water is taken out to irrigate valley stretches until the White Cañon in the Tewan Mountains is reached and the river rolls through a deep, rocky gorge for more than forty miles.[12] Emerging, its waters are again taken out upon the land from point to point until the entire territory is traversed, and the river passes out of New Mexico and becomes the boundary line between Texas and Mexico. From its source to the mouth of the Chama above the White Cañon it is a clear, deep river; below, it is a shallow river of mud. In this valley irrigation was practiced by the aboriginal village Indians

---

[10] Disputes over the interregional, interstate, and international apportionment of water led to negotiation of numerous multistate agreements dividing the water of western rivers (including the Colorado River Compact in 1922) and a treaty with Mexico in 1944 principally addressing the Colorado River and Rio Grande. Neither Powell nor the eventual signers of the allocation agreements contemplated the idea of leaving water in a river for purposes other than diversion and economic use.

[11] The San Luís Valley of south central Colorado.

[12] White Rock Canyon lies downstream of New Mexico's Española Valley, at the foot of the Jemez Mountains, which Powell calls "Tewan" in reference to the valley's Tewa Indian Pueblos. The Chama River enters the Rio Grande just above Española near San Juan Pueblo.

centuries before the discovery of America. Prior to 1600 it was populated by Spanish peoples coming up from Mexico.[13] So the gardens and
fields of the territory and the region along the river from El Paso to the
Gulf are old. Since the acquisition of the territory by the Government of
the United States irrigation has greatly developed in Colorado and New
Mexico, along the river itself in part, but mainly on the tributaries. No
waters have yet been stored in reservoirs, but the seasonal flow in dry
years is now wholly utilized; and more: the river for hundreds of miles
along its lower course is entirely cut off from a supply, and the gardens
and farms are now lying desolate and the winds are drifting the sands
over vineyard and field. During the past year more new works have been
projected than now exist in the valley. How are they to be supplied in
scant years? Who owns the water? Shall the men of Colorado take all
that falls in their State? and if so, shall the settlements in the valley of the
Rio Grande be destroyed by the new settlements on the tributaries? Just
across the line of New Mexico the town of El Paso, in Texas, is found;
and the town of Juarez lies on the opposite side of the river, in Mexico.
Here large areas have been irrigated and many thousand people are
engaged in agriculture; but they had little water last year, and the next
dry season they will have none. Shall the people who have cultivated the
land for more than a century be driven away?[14]

The Green River heads in the Wind River Mountains, and, rolling over
elevated cold plains, it at last reaches the Uinta Mountains, and plunges
through cañons to the mouth of the Grand. At its source the Grand inoscu-
lates with the Arkansas and the Rio Grande del Norte[15] and rolls through a

---

[13]Don Juan de Oñate established the first Spanish colony in New Mexico at the confluence
of the Chama River and Rio Grande in 1598.

[14]Powell testified at length to Congress on this subject. See "Ceding the Arid Lands to the
States and Territories." The Rio Grande had previously supported between 60,000 and
75,000 acres of agriculture on both sides of the river in the vicinity of today's El Paso and
Ciudad Juarez. Increased upstream diversions so reduced the quantity of available water
that at least half the irrigated area had gone out of production by 1890.

[15]Its headwaters interfinger with the headwaters of the other rivers. As mentioned earlier, the
name *Colorado* today also applies to Powell's Grand River. The junction of the Green and
the Grand lies downstream of Moab, Utah, in what is now Canyonlands National Park.

succession of cañons to the Green. Then the two rivers, joining in wedlock, become one indeed, and assume a new name, the Colorado of the West, which rolls into the Gulf of California. Its way for nearly 50 miles is through a succession of deep cañons, where it flows from 100 to 6000 feet below the general surface of the land. At last it emerges from the gloom of the Grand Cañon and runs in a valley through the lower portion of its course, now and then interrupted by a low range of volcanic mountains, through which it cuts its way in deep, black gorges. The region drained by the cañon portion of the Colorado and its tributaries and the region drained by the Grand and Green and their affluents are in the main inhospitable. All the streams flow through deep cañons between great blocks of naked rock, which are plateaus with cliff escarpments. Sometimes cañons widen into narrow valleys, and others are found at the foot of the mountains on the east and west, while far to the north are broad valleys inclosed by mountains; but these are cold and desolate. Some agriculture can be practiced by means of irrigation in the broad cold valleys above and the narrow warm valleys below, but a very small portion of the water of the Colorado will thus be used. A mighty river will ever flow from the mouth of the Grand Cañon. The region below the cañon on each side of the Colorado is one of great aridity, with an annual rainfall of not more than three or four inches. It is also a region of high temperature in summer, and it has almost a frostless winter. Here date palms flourish with a luxuriance never known in Egypt. Oranges, lemons, pomegranates, and figs grow and bear in abundance, and the lands are well adapted to sugar and cotton. On the west lie Nevada and California. On the east Arizona stretches away to the summit of the Rocky Mountains. The lands to which the waters can be taken greatly exceed the area that can be served. How shall they be divided? The low flood-plain along the river is narrow, and only small tracts within it can be redeemed. If the waters are to be used, great works must be constructed costing millions of dollars, and then ultimately a region of country can be irrigated larger than was ever cultivated along the Nile, and all the products of Egypt will flourish therein.[16]

---

[16] In 1901, little more than a decade after this article appeared, the Colorado Development Company brought the water of the Colorado River to the Salton Sink, a vast desert basin the

The northern third of Arizona is a lofty table-land; the southern part is a stretch of desert valley over which desert mountains rise. The descent from the table-lands to the lowlands is marvelously abrupt, for it is marked by a line of cliffs, the escarpment of a geologic fault. Along this fault there is a fracture in the rocks below, and the table-land side has been uplifted several thousand feet. Through the fissure of the fracture lavas have poured in some places, so that here and there the escarpment is masked by volcanic rocks. All of the perennial streams of the territory, that run to join the sea, head on the table-land or in the Rocky Mountains of New Mexico. The rainfall of the lowlands is insufficient to create ever-living waters. The land has never a carpet of verdure, but a few scattered desert plants are found, many of which belong to the cactus family. Everywhere the landscape is weird and strange. Most of the mountains are naked of vegetation or bear dwarfed gnarled trees of pine and cedar, with aloe[17] and cactus. The flood-waters that pour down these mountains sweep the disintegrated rocks into the valley below, and much of the region is filled to a considerable depth with sand and gravel. The storm-waters that come from the mountains sink into these valley sands and disappear; and the problem of this country is to gather the mountain waters into reservoirs at the foothills, and to recover them from the sands by artesian wells and pumps.[18]

In southern California there is another drop of the land from the San Bernardino Mountains to the coast, but its line is not so clearly marked as

---

[16] *(continued)* company renamed Imperial Valley. The further development of Imperial Valley, where today nearly half a million irrigated acres produce a large portion of the nation's winter vegetables, became closely intertwined with the hydraulic development of much of southern California, including the construction of Hoover Dam in the 1930s, a "great work" that was probably far greater than anything Powell contemplated.

[17] Probably a reference to yucca.

[18] Soon after its establishment in 1902 by passage of the Newlands Act, the federal Reclamation Service undertook to dam the Salt River upstream of Phoenix, much as Powell here advises. The dam, completed in 1911, and the resulting reservoir were named for Theodore Roosevelt, champion of the Progressive conservation movement, of which reclamation was a centerpiece.

that of Arizona. From this southward to San Diego and from the coast eastward but a few score miles there is a land of beauty. It is forever fanned with mild breezes from the Pacific, and thus cooled in summer and warmed in winter. When the rainy season comes its billowing hills are covered with green, and when the dry season comes the hills are covered with gold. The rainfall is almost sufficient for agricultural purposes; springs burst from the hills, and creeks meander to the sea. The little valleys open into broader marshes near the shore that are hardly above the tide, but they are often leveed by the waves of the sea, and wave-formed embankments beat back the high tides and protect the meadows that are inclosed by hills. Among the hills natural basins abound, into which the clouds may be enticed as they fall upon the ground, and into which the fountains may pour their waters. It is a region of country singularly well adapted to lakelet-reservoirs, where every man may construct one or more on his own farm. Little artificial supply is needed, and this can be easily secured; and a region of country about the size of Italy, with the climate of Italy, is rapidly becoming covered with the gardens of Italy.[19]

The Sierra Nevada culminates in altitude near its eastern margin. It is a great plateau declining westward, and carved into transverse ridges and valleys, that extend from the high eastern summit of the system to the low warm valley of California. Between the valley and the sea the Coast Range rises. The San Joaquin River heads in the heights of the south, and runs northward. The Sacramento heads far to the northward, where volcanic mountains stand. The rains and snows that fall on these peaks sink away into the scoria and sands of volcanic cones, and the mountains where the clouds gather and the storms rage are yet streamless; but away from the mountains, where volcanic sands disappear, the mountain waters burst out in mammoth springs, and creeks and rivers are born full grown. The Sacramento and the San Joaquin unite to flow through the Golden Gate.

---

[19] Powell rightly assessed southern California's Mediterranean climate. The region's hydraulic future, however, has depended far more on "great works" than on "lakelet" reservoirs.

In the southern or San Joaquin valley irrigation is already practiced, and the streams are partly or wholly used during the season of growing crops. The chief development of the area of agricultural lands in this region is to come from the construction of reservoirs for river and storm waters, and through the development of drainage systems, so that the water may be compelled to do double or treble duty. In the Sacramento valley irrigation has been practiced to a very limited extent, for the rainfall is considerable, and the people until the last year or two have been proud to affirm that their climate was humid; but they are now beginning to learn that even with them irrigation is highly advantageous, and that the product of the field may be multiplied more than threefold through the agency of rivers.[20]

It is in the valley of the Sacramento and its tributaries that the great deposits of gold gravels are chiefly found, and that extensive hydraulic mining has been carried on. The rivers of the Sierras were turned into reservoirs, and their waters, under high pressure, through the agency of monitors, were set to tearing down the hills of gravel and washing them away into the Sacramento. But these operations soon choked the stream and caused it to overflow the adjacent lands, and the sands and gravel brought down were deposited over the lands, and thus fields and towns were buried and populous regions were temporarily destroyed. Then the farmers of the valleys, through the legislature and the courts, stopped the mining operations; but strife still rages. The greed for gold and the hunger for fruit and wheat still spur the miners and farmers, and the conflict is irrepressible. Some day or other, when the madness has subsided, they will quietly discover that both parties are equally interested in the control of the rivers; that all of the waters of these regions can be stored in reservoirs and used at will, and that the valley of the Sacramento can be irrigated to multiply its agricultural products and its gold mines worked by the same agency, and that the miners and the farmers have common and harmonious interests in the hydraulic problems of the fairest land under the sun.

---

[20] Today irrigated rice farms are common in the Sacramento Valley.

In geologic times, not long ago as speaks the scientific man, but very long ago indeed as speaks the chronicler of human follies, there was a deep valley on the eastern slope of the Sierra Nevada at the headwaters of the Truckee River. About this valley towered granite mountains. But earthquakes came, and rents were formed in the rocks, and out of the fissures poured monstrous streams of lava. One of these fissures crossed the lower end of our mountain valley, and through it poured floods of molten rock. Stream after stream issued, to cool in solid sheets and blocks, until a wall was built across the valley two or three thousand feet in height, and above it was a deep basin five or six hundred square miles in area. The storms that fell on the granite and volcanic mountains rolled in rivers to fill the basin, and Lake Tahoe was created.[21] When filled, at last, its waters overflowed the rim of lava, and the Truckee River now springs from the Tahoe fountain. Its deep waters are dark with profundity, like the clouds of a stormy sky, but about its shores a few shallow bays are found, and emerald waters, like festoons of beauty, encircle the deeper and more somber lake. Back from the waters forest-clad slopes rise towards the heavens, and above are seen naked crags and domes of granite. Farther to the north, Donner, Independence, and other mountain lakes discharge their waters into creeks that join the Truckee. It is thus that a large hydrographic basin is formed in the mountains where torrential rains fall and deep snows accumulate in winter months, and in which the waters are collected to form the Truckee, which leaves the mountains in a dance of delight and with a never-ending song of laughing waters. Sweet valleys are found below, for the people have in many places reclaimed the desert and encircled their homes with verdant fields. But the waters are all caught in California, while the irrigated lands are in Nevada; so the farmers of the Saver State must go to the lands of the Golden State to construct their reservoirs. The water of the lake can be partly discharged each year by deepening its outlet and the water used for irrigation in

---

[21] Lake Tahoe is now thought to be a graben lake, formed by tectonic movement in the earth's crust that created a trough or depression at the surface. Lakes Baikal, Tanganyka, and Nevada's Pyramid Lake are also graben lakes.

Nevada, and after the irrigating season is over the gates may be closed and the lake permitted to refill; but this perhaps will mar a pleasure resort. Who shall judge between the States? A very large part of all perennial waters to be used in Nevada have their sources in California. Who shall judge between the States?[22]

In southern Utah a bold escarpment or cliff of rocks two thousand feet in height is presented towards Arizona. This is the edge of a plateau which extends far northward into central Utah. It is cut in two by a river which heads a little back from the brink of the cliffs and runs to the north; and so, except at the very southern extremity, two plateaus are found, which unite between the head of the river and the verge of the cliffs. This one–two plateau lies high and is covered with great forests, where rains and snows fall in abundance, and the waters gather to form the Sevier River. Along its upper course and beside some of its tributaries there are small valleys that are high and cold; yet grass, rye, oats, and potatoes can be raised in the short summer. Forty miles from its source the river enters a deep cañon, and when it emerges a broad and beautiful valley is found. Down this the stream meanders, and then turns westward and vanishes in the sand. It is a lost river. Just above the sink and along the valley through which the river meanders there is good and abundant land—much more than the river will serve; and here the Mormon people, who have institutions and customs like nations of the Oriental world in more than one respect, cultivate the soil by irrigation in the same manner. There are lands above the central cañon and lands below; but the river cannot serve them all. The earliest settlements were below. Later settlements have been planted above, in the sub-arctic lands, and they are taking the waters away from the older towns and farms. And how is Justice to be rendered between these conflicting interests?

North of Mt. Nebo lies Utah Lake, which is fed by the Provo River and a number of beautiful creeks. About the lake and along the streams the peo-

---

[22] Powell provided his answers to this and similar questions in the third article of this series, "Institutions for the Arid Lands" (Selection 14).

ple are cultivating the land by irrigation. But the surplus water is still discharged into the lake, which constitutes a great reservoir. From the lower end of the lake the river Jordan flows on to the Dead Sea of Utah, the Great Salt Lake, on whose shore the Mormon Temple stands. Large areas in the valley are watered from the river. The Utah Lake divides a hydrographic basin. On the Provo and streams above there are favorable sites for reservoirs, and there are areas of land that can yet be irrigated; but if the waters are used in the upper valley they cannot be used along the banks of the Jordan. All increase of the irrigated area above will decrease the irrigated area below. Who shall divide the waters and relegate them to the best lands in the interest of the greatest number of people?

Bear River has its sources partly in Idaho and partly in Wyoming. Where its upper affluent creeks are assembled it runs northward across the Utah–Idaho line. At this point it expands into a broad sheet of water known as Bear Lake, which is divided into two nearly equal parts by the territorial line. The surface of the lake is about six thousand feet above the level of the sea. The river, after leaving the lake below, runs northward for a long course into Idaho, and then turns upon itself and recrosses the territorial line into Utah. The course of this great curve is through cañons and cañon valleys, but at two or three points the valleys expand so as to present small areas of irrigable land. In general, above the Utah line, the region drained is mountainous. From this point the river flows through a steadily expanding valley until it empties into Great Salt Lake. Now it is possible to use much of the water of this stream in the upper region on mountain valley lands, where hay can be cultivated and some other of the crops of cold climates. Another portion can be used in Idaho, while the great valley along the whole stretch of the river is admirably adapted to irrigation. Bear Lake itself, which lies in two Territories, is ultimately to become the chief reservoir, but others can be constructed above, and still others below. Thus the reservoir system must be distributed between the two political divisions, while the great body of the lands to be redeemed are in Utah. How these lands are to be selected, and water-rights relegated to such lands, is a serious problem which demands immediate solution, for the people are already in conflict. Angry passions have been kindled, and war would ensue were it an international instead of an interstate problem.

The Snake or Shoshone River heads in the great forest-clad mountains of Wyoming and runs across the line into Idaho, then passes quite across the Territory until it becomes the boundary line between Idaho and Oregon. Passing the northeastern corner of the last mentioned State, it enters the State of Washington, and runs westward for a long reach until it debouches into the Columbia. The Shoshone River is one of great volume, second only to the Colorado. Reservoir sites along its course in Wyoming and Idaho have already been revealed by the surveys, and it is shown that in the upper region water can be stored to an amount of more than 2,000,000 acre feet. This will irrigate at the first usage at least 2,000,000 acres of land; and if they be properly selected, so that the waters can be collected again and again after serving the land, the area redeemed will be more than 4,000,000 acres. There are many other tributaries below that have not yet been examined, and it is safe to say that the waters of the Shoshone with its tributaries may ultimately serve from 8,000,000 to 10,000,000 acres.[23] In its utilization three classes of problems are involved. If the waters are taken out in small canals near to the river and the lowlands served first, and prior rights and interests established on such lands, then but a small part of the stream can be used, and the greater part will run away to the Pacific Ocean; and subsequently the region of irrigation can be enlarged only by buying out vested water-rights scattered along the course of the river. But if at the very beginning the water can be taken out high up the river and carried in great canals to either side and there distributed to the higher lands, and used over and over again on its return, a complete utilization can be secured, and the cost of the construction of the system of irrigation by

---

[23] In 1997 about 3,500,000 acres were under irrigation (including groundwater irrigation) in Idaho, nearly all of it on the Snake River Plain in the south of the state. It is hard to tell whether Powell also means to include the lands of Washington State in his estimates, but in 1997 only 1,705,025 acres were irrigated from all sources in Washington.

reservoirs and canals will be greatly reduced per acre. To irrigate 2,000,000 acres of land near to the river by short canals taken out along its course here and there will cost more than half as much as the construction of hydraulic works that will serve from 6,000,000 to 8,000,000; while the scattered minor works will be forever subject to destruction by the floods, and the agriculture secured will be of less value per acre, because the best lands will not be served, and only imperfect drainage will be secured.

The valley of the Shoshone has an interesting structure. In late geologic times it has been the site of great volcanic activity. The eruptions have not produced cones and mountains, but fissures have been opened and broad sheets of lava have been poured out over the region. It is a valley of volcanic mesas or low table-lands. On the basaltic rocks thus poured out a peculiar surface is developed. The floods of cooling lava roll down in waves and bubble up in domes, which often crumble and fall in, leaving many pits, and the general surface is thus exceedingly irregular; but the irregularities are not on a great scale so as to produce high hills and mountains. The process of degradation by frost and heat, by wind and rain, smooth out these irregularities; the higher points are degraded and the lower places are filled. Many of the eruptions in this valley are of such age that their surface has been smoothed out in this manner; but there are many others so irregular that the mesas are covered with pits and naked rocks, and are thus wholly worthless for agricultural purposes. The second great problem is properly to select the mesa lands to which the waters shall be distributed. A part of the storage of the water must be in Wyoming, while the lands to be served must be in Idaho, Oregon, and Washington. These are interests over which nations would speedily be at war; in this country they involve interstate questions, and must be settled by the General Government.

Space fails me to describe the beautiful lands of the Columbia and its tributaries, but interstate and international problems are involved. The Columbia comes from British territory. One of its affluents, the Kootenay, heads in British territory, passes into Montana, and returns to British territory. Passing over to the Missouri, some of its waters head in foreign lands, and Montana, North Dakota, South Dakota, and Nebraska are interested.

Along the hundredth meridian from Manitoba to Mexico there is a zone of semiarid land.[24] Years ago, when the writer first began investigations into the agricultural prospects of the far West, he abandoned the designation "desert" and adopted the term "arid," as more properly characteristic of the country. For the one hundredth meridian zone he at first adopted the term "sub-arid," but it gave great offense, and the suggestion that irrigation was necessary to its successful cultivation was received with denial and denunciation, for at that time the advantage of artificially supplying water to cultivated lands was generally unknown. Seeing that the term "sub-arid" was a red flag to kindle anger, it was dropped, and the term "sub-humid" was adopted; and now the hundredth meridian zone is generally known as the "sub-humid" region. The average rainfall, which varies much from year to year, is about eighteen inches on its western margin, and increases to about twenty-four on its eastern edge. Passing from east to west across this belt a wonderful transformation is observed. On the east a luxuriant growth of grass is seen, and the gaudy flowers of the order Compositae[25] make the prairie landscape beautiful. Passing westward, species after species of luxuriant grass and brilliant flowering plants disappear; the ground gradually becomes naked, with "bunch" grasses here and there; now and then a thorny cactus is seen, and the yucca thrusts out its sharp bayonets. At the western margin of the zone the arid lands proper are reached. The winds, in their grand system of circulation from west to east, climb the western slope of the Rocky Mountains, and as they rise they are relieved of pressure and lose their specific heat, and at the same time discharge their moisture, and so the mountains are covered with snow. The winds thus dried roll down the eastern slope into lower altitudes, when the pressure increases and they are heated again. But now they are dry. Thus it is that hot, dry winds come, now and then, and here and there, to devastate the sub-humid lands, searing the vegetation and parching the soil. From causes not well understood

---

[24] The eastern boundary of the Texas panhandle follows the 100th meridian, which roughly bisects Nebraska and the Dakotas.

[25] This large family of plants includes sunflowers, daisies, asters, and other familiar genera.

the rainfall often descends in fierce torrents. So storms and siroccos alternately play over the land. Here critical climatic conditions prevail. In seasons of plenteous rain rich crops can be raised without irrigation. In seasons of drought the fields are desert. It is thus that irrigation, not always a necessity, is still an absolute condition of continued prosperity. The rainfall is almost sufficient, and the artificial supply needed is small—perhaps the crop will rarely need more than one irrigation. A small supply for this can be obtained from the sands of the river valleys that cross the belt. In some regions artesian waters are abundant; but the great supply must come from the storage of storm-waters. The hills and mesas of the region are well adapted to this end. Under such conditions farming cannot be carried on in large continuous tracts.

Small areas, dependent on wells, sand-fountains, and ponds, must be cultivated. It is a region of country adapted to gardens, vineyards, and orchards. The hardier fruits can be cultivated at the north, and sub-tropical fruits at the south. From this region the towns and cities of the great valley and the capitals of trade in the East will be supplied with fruit and vegetables. It is the region of irrigation nearest to them, where gardens and fields produce richer, sweeter products than those of humid lands. Already the people are coming to a knowledge of this fact and are turning their industries in the right direction. The earliest settlements have been planted in seasons of maximum of rain [sic], and the people who came had dreams of wealth to be gathered from vast wheatfields. Now wholesale farming is almost wholly abandoned. In the last twenty years, during which the writer has been familiar with the sub-humid zone, having crossed it many times and traversed it in many ways, he has seen in different portions two or three tides of emigration, each ultimately disastrous, wholly or in part, and settled regions have become unsettled by migration to other districts. But from each inflow a few wiser men have remained and conquered prosperity; and now that the conditions of success are known, he is willing to prophesy—not from occult wisdom, but from a basis of fact—that the sub-humid region will soon become prosperous and wealthy.

The Arid Land is a vast region. Its mountains gleam in crystal rime, its forests are stately, and its valleys are beautiful; its cañons are made glad with the music of falling waters, its skies are clear, its air is salubrious, and it is already the home of millions of the most energetic men the world has ever known.

———◇———

The Non-Irrigable Lands of the Arid Region

———

Here Powell addresses the character and use of rangelands and forests, the lands that today make up most of the public lands in the intermountain West.

———

Sun is the father of Cloud.

Cloud is the mother of Rain.

Sun is the ruler of Wind.

Wind is the ruler of Rain.

Fire is the enemy of Forest.

Water is the enemy of Fire.

Wind feeds Forest, and Rain gives it drink.

Wind joins with Fire to destroy Forest.

Constant Rain battles with fickle Wind and mad Fire to protect Forest.

So Climate decks the land with Forest.

There are very large areas of the world unclad with forests, but this is not for lack of rain. Forests, low, gnarled, thorny, and scant, will grow with even less than ten inches of annual precipitation. Such are the forests of sunny Ari-

———

Century Magazine 39 (April 1890): 915–22.

zona. As the rainfall increases from clime to clime, the forests become more luxuriant, stately, and dense, until with sixty inches of rainfall a growth is produced which almost baffles description. Then giants crowd one another and lift their heads higher and higher in rivalry to bathe their verdant crowns in sunlight. High and straight towards the heavens they thrust their boles, and their boughs push towards the zenith by the shortest way of verticality. The young trees also are slender and straight, and depend on the giants for protection against overthrowing blasts. Around the feet of the giants is a dense undergrowth. But old trees die and fall, and their great stems lie on the ground or are held above it by large branches. Through this warp of living and dead trees there is a woof of vines, climbing the trees, running out on the branches, creeping over logs, and stretching from tree to tree, branch to branch, and log to log, all woven into a mass of vegetation. Thus the erect and creeping living and the prone and prostrate dead constitute a forest tangle into which man can penetrate only with the greatest toil. Such are the forests that stand about stormy Puget Sound.

Between these extremes there are many degrees of luxuriance in tree growth. When a region is reached with less than forty inches of rainfall small prairies may sometimes be found, and passing on to regions of still less rain the prairies are larger and more frequent. When districts of about thirty inches of rainfall are reached prairie predominates, and the few and smaller forests are called groves. Still passing to zones of less precipitation the prairies become plains, and such forest growth as may be found is mainly ranged along the river banks or scattered over stony hills.

If there were no intervening agency, climate would cover the earth with trees wherever there is more than ten inches of rain. This agency is fire. Rainfall, then, furnishes the potential limit to forest growth, fire the actual limit. On the other hand, rainfall furnishes a limit to fire in such a manner that it becomes less and less destructive, until, under mean conditions of latitude and altitude, forty inches of yearly rain establishes a practical limit to its ravages. In a region where prairie and grove divide the land between them, fire and storm are evenly matched. Fire is king on the plains; storm rules where the forest stands.

The arid lands of the United States are chiefly without trees, although the rainfall is sufficient for their production except in desert areas of Arizona and California; but fire prevents their development or destroys them after they are grown. Still, some areas of the country are wooded. Along the streams grow cottonwoods of value for firewood and for minor domestic purposes. On elevated mesas or table-lands, and on lofty hills, are scant forests, consisting mainly of low, straggling piñons, or nut pines, dwarfed and gnarly cedars, and ragged and deformed oaks. These forests do not furnish milling timber, but they are useful for fuel and for many other purposes. On the higher plateaus and mountains great forests are found, composed of pines of many species, spruces, hemlocks, firs, and sequoias. The timber trees are all coniferous and needle-leaved. The oaks are but bushes, often Lilliputian. Some of the oaks of arid Texas vainly vie with the goldenrods of Illinois; while the cactus plants of the Prairie State would look up with wonder to the cactus plants of Arizona, as pygmies gaze on giants. The oaks of the foothills along the western slope of the Sierras in California attain a greater size, and become orchards of acorns, where Indian hunters and grizzly bears were wont to compete for food in the days when the soil was unscarred by the miner's pick. The forests of the plateaus are not dense, though the trees are stately, and the lands are often variegated with brilliant chaparral and blooming prairie.

The mountains are not uniformly clothed with woods, but here is a grove of pine, there one of spruce, hemlock, or fir. Often these trees are commingled, and in the Sierras of California sequoias stand above them all. By the streams and in the mountain glades silver-stemmed aspens abound, whose wealth of foliage turns to gold when the autumn rime appears. Sometimes a driving wind sweeps through such an aspen grove and brushes the leaves from their twigs, and they float on the air like a cloud of butterflies, resplendent in the brilliant sun of a cloudless sky. Many a mountain side is naked, and many a peak is lifted above the timber line into the region of snow and ice.

We mount our horses at Flagstaff in northern Arizona. In ten minutes we are in the woods and out of sight of the railroad town. We ride for hours among the pines, and from time to time see San Francisco Mountain on our right. Here and there, as we go, a black cinder-cone is lifted for a few hundred feet, aspen groves are seen, and at noon we ride up the slope of a low, dead volcano, and, passing a rim of crunching cinder, halt on the shore of a lakelet in a crater. Then on we ride through an open pine forest, until at last we come down to hills that are covered with piñons and cedars, and rest for the night by a spring concealed among oak bushes. It has been a long ride, and we sleep well. Before the morning sun illumes the hilltop we are on our way again—still to the north, across sage-brush plains and cedar-clad hills; by noon we are once more on the verge of a pine forest, and we lunch by a water-pocket that was filled by a storm two months ago.[1] Then our way is across glades carpeted with flowers, and through open forests where we now and then see a deer bounding on its way. So we pass over prairie and through pine forest until at last we reach the brink of the Grand Cañon of the Colorado. When the days of wonder-seeing are past, we turn to the southwest, riding through forest and across prairie. At intervals of twenty or thirty miles we find a spring or a water-pocket. And so we journey, day by day and week by week, over prairies, through forests, and among cinder-cones and dead volcanoes, glad to find a water-pocket after a long ride and supremely happy to camp by a living spring. But no creek, no river, is ever found. Such is one of the great forest-clad plateaus of the arid region.[2]

Our steeds are now psychic, and we amble through air to Middle Park in Colorado, and camp at the foot of a mountain. Near by rolls Grand River, and there by the rock is a fountain whose waters come from unknown depths, where they have been heated in the caldron of eternal fire.[3] From the boiling waters a cloud of steam arises, loaded with sulphurous odors, and a pellucid brook flows over a carpet of brilliant *confervae*[4] on its way to

---

[1] A water pocket, known also in the Southwest by the Spanish term *tinaja*, is a rock basin that captures snowmelt and rain.
[2] Powell is describing the Coconino Plateau and other tablelands of central Arizona.
[3] Probably Hot Sulphur Springs, near the contemporary town of the same name in Grand County, Colorado, southwest of Rocky Mountain National Park.
[4] *Confervae* is a filamentous green alga of the genus Tribonema; it grows in still or slow-moving water.

the river. When morning comes again we continue our ride on terra firma, among hills and then among mountains. Now and then we come to a stream where our horses must swim, and we wade creeks and leap over brooks until we plunge again into forests beset with fallen timber.

At noon we camp on the margin of Grand Lake,[5] here bordered with stately forests, there walled with precipitous rocks. True, the distance is great for a morning ride, but our chargers are the best—why not? They are imagination-bought, and we have wealth of fancy. For the afternoon we plunge into a dead forest where a fire played havoc ten years ago. Some trees are prostrate and obstruct the way. Falling trees have caught in the branches of those still standing, and lean here and there with varied angles. Trees supported by others, trees prostrate and trees erect, naked white trees with naked white arms, are woven into a maze of ghostly bars to block our way. Over and under and around we pursue our course. Then a storm comes on. The wind sweeps through this ancient battlefield of fire and storm, and the stark, dead limbs crack, break, and crash on the ground. Now and then a great stricken tree falls and fills the air with a roar which vies with the thunder. Dead trees caught in the arms of dead trees sway and shriek, and the tempest runs mad with wild delight. We stand on open ground and gaze on the destruction and listen to the battle-music of nature. When the storm has passed we ride along until live woods are reached, and at night camp where a mountain rill lulls us to sleep. So for days and weeks we ride through dead forests and live forests, and everywhere in the mountains we find rivers, creeks, brooks, springs, and lakes. Such are the forests of the Rocky Mountains.

Once more, on steeds as swift as dancing light, we enter a grove of live-oaks in the valley of California. Where other trees have curves, these have angles; they are all knees and elbows, and they stand akimbo with knotted fists. But, as if to hide deformity, they are covered with a mantle of perennial green. Now we ride over meadows of green and hills of gold until more symmetric oaks and cedars are found; blue pines are seen, and at night we

---

[5] Grand Lake is located at the western boundary of Rocky Mountain National Park.

reach the great sugar pines of the Sierra. Then we slowly climb the long, gentle slope to the west. Cedars like those of Lebanon on every hand, pines like those on Norwegian hills, and at last we see a sequoia, the grandfather of trees. Past the big trees, we next day find forest and chaparral contending for the land. The woods are of pines and spruces and firs, and the chaparral is brilliant with the scarlet boughs of manzanita and gnarled mountain mahogany. High up the mountain we climb, and the pines are lost, the spruces disappear, and the firs are dwarfed, until we are among domes of gray granite and pinnacles of trachyte,[6] and down into a vast amphitheater of sheer rock comes a creeping glacier.[7] So on we ride from day to day, week to week, and month to month, from dwarfed fir above to dwarfed oak below, and again from foothill to granite dome, until we have crossed all the rivers that flow from the Sierras and unite to pass through the Golden Gate. During this ride we have seen the great Sierra forest.

For a number of years a survey of the arid lands has been in progress, and the forest areas have been mapped, and they have all been studied more or less. Now surveys are mathematical, for relations of quantity are involved. Numbers perhaps are more arid than land, and hence they are appropriate here. Glance at the following table, and some idea will be obtained of the comparative extent of the forests of which I have spoken.

It will appear from the above table that about one-tenth of the arid region is covered with firewood timber, but this timber is very scant, and often the open spaces are large. It could all stand on one-fiftieth of the entire arid area and not be crowded. The milling timber also covers about another tenth of the ground, but there are many barren places, and usually the trees are widely scattered, so that they could all stand on one-fortieth of the space

---

[6] Trachyte is a light-colored, fine-grained igneous rock.

[7] Powell would have been aware of Clarence King's 1871 paper, "On the Discovery of Actual Glaciers on the Mountains of the Pacific Slope" in *The American Journal of Science and Arts.* A 1972 survey found 497 glaciers and 847 ice patches in the Sierra Nevada. The glaciers, which have a cumulative area of less than 20 square miles, are not remnants of the great glaciers of the Pleistocene; they are relicts of the Little Ice Age of roughly 1700–1750. (See Hill, *California Landscape: Origin and Evolution.*)

Approximate Area in Square Miles of Timbered Lands in the Arid Region[8]

State	Firewood (square miles)	Merchantable Timber (square miles)
Washington	1,050	1,080
Idaho	8,600	9,800
Montana	6,500	21,000
Oregon	3,500	8,700
Wyoming	7,300	15,700
South Dakota	2,400	400
North Dakota (river bottoms)	200	—
California	20,300	11,000
Nevada	5,400	700
Arizona	26,510	11,700
New Mexico	21,540	14,490
Colorado	15,000	23,500
Utah	14,000	7,700
Totals	132,300	125,770
Grand total	258,070	

Total area of arid lands	1,331,151 square miles

and still have abundant room. So both classes combined could easily stand on less than one-twentieth of the arid region.[9]

The merchantable timber is all on the high plateaus and mountains; hence the lands where it grows are not valuable for agricultural purposes. Cañon walls, cliffs, crags, and rocky steeps are not attractive farming-

---

[8] Lest the small area of merchantable timber attributed to Washington, Oregon, and California appear surprising, bear in mind that the great temperate rainforests of the Pacific Northwest are excluded from these calculations.

[9] This is a puzzling analysis. Powell seems to be saying that both classes of forest are exceedingly patchy—interspersed with glades, parks, and other open spaces—and that if they were more uniform, they would take up less space. In a sense, he is describing a mapping difficulty as a fault of the forests: His maps would be better if the forests were more compact. He may also be misunderstanding the ecological needs of forests in which trees are widely spaced in order that each have sufficient water, nutrients, and, in some cases, sunlight.

grounds. But more: at these great altitudes deep snows fall, ice appears early and lingers long, and frosts come on many a summer night.

The agricultural lands are situate in the valleys where the streams flow. Thus forest and farm are dissevered by dozens and scores of miles. So forest industries are segregated in one region, farming industries in another. It is no small task for the farmer and the villager to haul their wood from distant mountains and to bring poles and logs from the upper region, for it is a day's or a week's journey, and roads must be made over hills and along mountain sides. In many places flumes are constructed—great canals in lumber troughs that stand on trestles, into which creeks are turned, and the lumber is floated down to the habitations of man. Then railroads and tramways are constructed for the same purpose. Often "slides" are built, by arranging two parallel lines of logs down the mountain side, between which the timber glides. It is thus that the valleys are dependent on the mountains through the agency of a special lumber industry.

The miners are also interested in these forests. As they penetrate with their shafts, drifts, and galleries into the hills and mountains, they carry away to the surface the rock in which the gold, silver, copper, and lead are found, that the metals may be extracted on the ground above. Then they are compelled to support the overhanging walls, that they may not crumble down. When great depths are reached, the enormous weight of superincumbent mountain squeezes the floors of these galleries and causes them to creep up. To prevent crumbling from above and creeping from below the underground spaces are densely propped with timbers; so thousands and millions of cords of wood are used underground. The forests are also valuable for fuel in metallurgic processes, and to furnish the power necessary for running mining machinery. Many of these mines are in the mountains, and the timber grows near by; sometimes it grows far away, and must be hauled or transported by rail or flume to the mines where it is needed. So the mining operations largely depend on the forests.

More than two decades ago I was camped in a forest of the Rocky Mountains. The night was arched with the gloom of snow-cloud; so I kindled a fire at the trunk of a great pine, and in the chill of the evening gazed at its welcome flame. Soon I saw it mount, climbing the trunk, crawling out along the branches, igniting the rough bark, kindling the cones, and setting fire to the needles, until in a few minutes the great forest pine was all one pyramid of flame, which illumined a temple in the wilderness domed by a starless night. Sparks and flakes of fire were borne by the wind to other trees, and the forest was ablaze. On it spread, and the lingering storm came not to extinguish it. Gradually the crackling and roaring of the fire became terrific. Limbs fell with a crash, trees tottered and were thrown prostrate; the noise of falling timber was echoed from rocks and cliffs; and here, there, everywhere, rolling clouds of smoke were starred with burning cinders. On it swept for miles and scores of miles, from day to day, until more timber was destroyed than has been used by the people of Colorado for the last ten years.[10]

I have witnessed more than a dozen fires in Colorado, each one of

---

[10] Powell's fondness for telling this story did not endear him to forest preservationists. He related it, evidently less concisely than he does here, at a meeting with secretary of interior John Noble on December 30, 1890, much to the dismay of a group of forest advocates who included Bernard Fernow, then the head of the U.S. Department of Agriculture's fledgling Division of Forestry. Fernow and the others had organized the meeting to press Noble to close headwater areas in the public domain from public entry. Wrote Fernow, "Major Powell launched into a long dissertation to show that the claim of the favorable influence of forest cover on water flow or climate was untenable, that the best thing to do for the Rocky Mountains was to burn them down, and he related with great gusto how he himself had started a fire that swept over a thousand square miles. He had used up our time when our chance came to speak." (As quoted by Pisani, "Forests and Reclamation, 1891–1911," 70.) Fernow's successor, Gifford Pinchot, similarly derided Powell's account of the fire he started, although where Fernow was prejudicial, Pinchot was merely belittling: "Major John Wesley Powell . . . told in print how he had set fire to a great alpine tree just to see it burn. His account of how that fire caught other trees and then went roaring off through the Western mountains gave no slightest indication of regret. It was interesting, and that was all" (Pinchot, *Breaking New Ground,* 24). In fact, it is much more than interesting. Powell is here arguing, in a manner exceptional for his time, that fire is among the primary forces shaping the forests of the West and that not all of its influences are injurious to human interests. This fundamental ecological truth entirely escaped the attention of Pinchot, Fernow, and other "right thinking" conservationists of the Progressive era, for whom fire was an entirely evil force, a wasteful intruder upsetting to the "balance" of nature.

which was like that described. Compared with the trees destroyed by fire, those used by man sink into insignificance. Some years ago I mapped the forests of Utah, and found that about one-half had been thus consumed since the occupation of the country by civilized man. So the fires rage, now here, now there, throughout the Rocky Mountains and through the Sierras and the Cascades. They are so frequent and of such vast proportions that the surveyors of the land who extend the system of triangulation over the mountains often find their work impeded or wholly obstructed by clouds of smoke. A haze of gloom envelops the mountain land and conceals from the eye every distant feature. Through it the rays of the sun can scarcely penetrate, and its dull red orb is powerless to illumine the landscape.

During last season I made a trip over the arid lands by rail. On the way through the Dakotas the landscape was covered with a veil through which it was as vain to peer as through a fog at sea. On we went, meandering through the cañons and among the great ranges of Montana, but the smoke covered all the landscape of mountain forms, and for aught that could be seen we might as well have been crossing featureless plains. Then we passed through Washington and Oregon and down through Idaho—ever in a mountain land, and never a mountain in sight. As we crossed the line into Utah a shower came and cleared the atmosphere, and behold! the Wasatch Mountains were in view; a great facade of storm-carved rocks beetled above the desert as proud as if they were not doomed to be destroyed by storms and buried low in the valleys by rivers.[11]

It is thus that, under conditions of civilization, the great forests of the

---

[11] Powell included this description in testimony to the House Committee on Irrigation, February 27, 1890, adding, "That has been the experience for twenty-odd years, year by year, in this region. The geographical work of our Survey is cut off during the very dry months by the smoke; the men can not get lines of sight from height to height because of the fires produced in the mountains and the smoke settling down over the land. In the last twenty years one-half of the timber has been burned." (See House Report 3767, "Ceding the Arid Lands to the States and Territories," 79.)

arid lands are being swept from the mountains and plateaus.[12] Before the white man came the natives systematically burned over the forest lands with each recurrent year as one of their great hunting economies. By this process little destruction of timber was accomplished; but, protected by civilized men, forests are rapidly disappearing. The needles, cones, and brush, together with the leaves of grass and shrubs below, accumulate when not burned annually. New deposits are made from year to year, until the ground is covered with a thick mantle of inflammable material. Then a spark is dropped, a fire is accidentally or purposely kindled, and the flames have abundant food.[13]

There is a practical method by which the forests can be preserved. All of the forest areas that are not dense have some value for pasturage purposes. Grasses grow well in the open grounds, and to some extent among the trees. If herds and flocks crop these grasses, and trample the leaves and cones into the ground, and make many trails through the woods, they destroy the conditions most favorable to the spread of fire.[14] But if the pasturage is crowded, the young growth is destroyed and the forests are not

---

[12] Here is an interesting departure from the analysis of the *Report on the Arid Lands*. Where previously Powell had attributed the prevalence of fire in Western mountains to Indian fire-setting (see p. 176), he now acknowledges that a veritable epidemic of wildfire accompanied white settlement and that the settlers themselves—miners, lumbermen, herdsmen, travelers, etc.—were sources of ignition.

[13] Powell probably discussed Indian burning practices in the meeting with Secretary Noble described in note 10, and if so, he probably cited the purposeful use of fire by the Paiute bands he had so closely studied. Historian Stephen J. Pyne suggests that the term *Paiute forestry* was born from this encounter. Fernow, Pinchot, and other advocates of all-out fire suppression (including the young Aldo Leopold) thereafter derogated the practice of low-intensity, "light" burning as Paiute forestry. Today ecologists and foresters recognize light fire as an essential tool in managing a wide range of forest, woodland, and grassland ecosystems, especially in the American West (see Pyne, *Fire in America*, 100 ff.). Powell rightly concludes that in the absence of frequent, light fires, fuels steadily accumulate, leading to hotter, more destructive fires.

[14] Again, Powell shows a strong grasp of ecological dynamics. Grazing removes the fine fuels on which periodic light fires depend, and trails and roads function as firebreaks. The cessation of naturally occurring light fires in western forests where they had previously been common generally coincides with the commencement of widespread grazing. (See Bogan et al., "Southwest," and Allen, *Fire Effects in Southwestern Forests*, 28, 41.)

properly replenished by a new generation of trees. The wooded grounds that are too dense for pasturage should be annually burned over at a time when the inflammable materials are not too dry, so that there may be no danger of great conflagration.[15]

The area of good timber being very small, it has great value, and its rapid destruction is a calamity that cannot well be overestimated. These living forests are always a delight, for in beauty and grandeur they are unexcelled; but dead forests present scenes of desolation that fill the soul with sadness. The vast destruction of values, together with the enormous ravishment of beauty, have for years enlisted the sympathy of intelligent men. Forestry organizations have been formed; conventions have been held; publicists have discussed the subject; and there is a universal sentiment in the West, and a growing opinion in the East, that measures should be taken by the General Government for the protection of the forests. This subject is of profound interest; but sometimes factitious reasons are given which detract from the argument for the preservation of the woods.[16]

In humid lands, where rivers flow on to the sea because they are not needed on the fields, the problems relating to the streams are of another character. There the floods are destructive, and every condition which favors their diminution is an advantage. Vegetation lives on water. The roots drink it, and the leaves return all that is unused to the air, where it may float away to form clouds in other regions. A vigorous plant will thus evaporate two or three hundred times the weight of its annual growth. Then a great

---

[15] Powell is acknowledging a main concern of forest protection advocates: that too great a concentration of livestock, especially sheep and goats, will prevent regeneration of desired trees. He has urged a managed approach to grazing where it is possible, not the unrestrained conditions of the open range. Where conditions prevent such an approach, he urges the thoughtful use of fire—altogether a visionary stance for his day.

[16] *Factitious* means "fictitious, false." Powell quite reasonably qualifies his support for forest protection with a critique of some of the erroneous arguments advanced by forest protectionists, chief among them the notion that forests make arid climates wetter and invariably boost water supply. As will be seen, he also remains opposed to permanent federal control of forest management, which was then a primary goal of most protection efforts.

tree spreads, through the agency of its leaves and branches, a vast surface to the air and the heat of the sun. Altogether no inconsiderable portion of the precipitation of a region is thus returned to the heavens, and so fails to find the rivers. The subject has been more or less studied, but it is complex, and the result cannot be simply stated, for the variables are many. Perhaps it is safe to say that from twenty to forty percent of the rainfall of a region may be dissipated in this manner. It is manifest that such a loss from the streams is of no small importance in a humid region, and it is for this reason that the preservation of mountain forests in such lands is often strongly urged. But when the streams have a value which increases with their volume, the economic aspect of the problem is at once reversed. Researches on this subject made in the Wasatch Mountains and elsewhere by scientific men show that a great increase in the volume of the streams may accrue from the denudation of the mountains of their evergreen garments. There is still another condition which tends in this same direction. When the mountain declivities are grassy slopes, the snows of winter drift behind ledges and cliffs and into great banks among the rocks, and they fill ravines and cañons, and are thus stored in compact bodies until they are melted by the summer suns and rains. But when forests stand on the slopes the snows are spread in comparatively thin sheets, and great surfaces of evaporation are presented to the sun and the wind. For all these reasons the forests of the upper regions are not advantageous to the people of the valleys, who depend on the streams for the fertilization of the farms.[17]

But there is an obverse side of this problem. When the waters are stored for irrigation in natural and in artificial lakes the preservation of their reservoirs is of prime importance. Storm waters wash the sands from naked hills and mountains, and bear them on to the creeks and rivers, by which they are carried to the storage basins. Protection from these injurious agencies is

---

[17] Trees do indeed transpire large amounts of water, and streamflow declines as the forests of a watershed grow excessively dense. Nevertheless, Powell somewhat overstates his case, much as did the forest protection advocates who argued the opposite—that forests uniformly increase water yield. See Selection 10, note 7.

chiefly afforded by vegetation. For this purpose grass and chaparral serve well, but woods are better.[18] For the protection of reservoirs, therefore, it is important that their immediate slopes should be forest-clad, and that all declivities above, the waters of which cannot be discharged in large part of their sediments before reaching the reservoirs, should also have their woods preserved. In the utilization of these timber regions, then, as a source for the lumber which the people need, judgment and circumspection will be necessary properly to select the areas to be denuded.[19] It is thus

---

[18] In general, Powell is right but exceptions exist. In the Southwest, for instance, grasslands afford better protection from erosion than some woodland stands, especially piñon and juniper.

[19] In 1888, Henry Gannett, one of Powell's geographers in the Geological Survey, published two articles rebutting the idea that forest growth prompts increased rainfall. His arguments were well reasoned and empirically based but led him to adopt an extreme position. In the latter article he concludes, "In our arid region, which is dependent for irrigation on its streams, it is advisable to cut away as rapidly as possible all the forests, especially upon the mountains, where most of the rain falls, in order that as much of the precipitation as possible may be collected in the streams. This will cause, not a decrease in the annual flow of streams, as commonly supposed, but an increase . . . rendering fertile a greater area of the arid region. It may be added that the forests in the arid region are thus disappearing with commendable rapidity." Some contemporary observers and later historians surmised that Powell's views and Gannett's were the same (Pisani, *To Reclaim a Divided West,* 161–63), but Powell is not so easily pinned down. Throughout his later years, Powell insisted that his position on forests was consistently misinterpreted, and indeed he lent support to a number of forest protection efforts, including the preservation of the Kaibab country north of the Grand Canyon.

Nevertheless, by pointing out the complex relationship between forests and water yield and by his consistent embrace of a utilitarian point of view, Powell invites multiple readings. The present selection offers a case in point. In observing that "the forests of the upper regions are not advantageous to the people of the valleys," some readers may conclude, probably wrongly, that Powell is implicitly endorsing the kind of systematic clearing of forest stands that Gannett proposed. Similarly, when Powell speaks of denuding mountain slopes, a close reading of the context suggests that he is referring to the kind of logging necessary to provide timber for settlements, yet one might also draw from his words a more expansive and generalized endorsement of forest clearing. These ambiguities resolve most satisfactorily when one recognizes Powell's reluctance to be prescriptive. His concluding point that "judgment and circumspection will be necessary properly to select the areas" where timber might be cut begs the question of whose judgment should be used. His answer comes in Selection 14, in which he forcefully argues that local communities should decide which lands to cut and which to leave untouched, for they alone would be obliged to live with the consequences.

that the people of the valleys are interested in the forests of the mountains. Among the crags and peaks where winter winds howl, and where the snows fall all winter long, there grow inchoate cottages and schoolhouses and the fuel that illumines the ingleside.[20] And the mountain passes are the portals through which the clouds of heaven come down to bless their gardens and their fields, and to fill the fountains from which their children quaff the water of life.

The lowlands of the arid region are dry and hot, and are almost destitute of grasses. The summits of the highest mountains are in regions of almost perpetual frost, and grasses are practically wanting. Between these extremes of mountain top and desert valley there are vast areas of nutritious grasses, scant below, but becoming more luxuriant as one climbs the hills, traverses the plateaus, and wanders over the mountain sides. The lowest lands, those bearing more scant grasses, are the lands to be irrigated, for the waters can be taken to them. The better pasturage lands are usually too high for agriculture.

Climatic temperature decreases from the level of the sea to the summit of the mountains, but it also grows colder from the equator to the poles. Now the lowest lands of the arid country are farthest south. In Arizona and southern California the uninhabitable deserts of America are found; there are districts of country below the level of the sea and other stretches just above it. These low, torrid lands are strewn with pebbles, over which the winds sweep and carry on their way a load of sand as an instrument by which the pebbles are polished. It is thus that the desert in many places is paved with a mosaic of gems that gleam in many colors and blind the eye with their radiance. There are other stretches where billows of sand drift across the desert with the prevailing winds. Still other areas are covered with sand and stony fragments and strewn rocks, where vegetation gains lit-

---

[20] Powell's poetic efforts here threaten incomprehensibility. The "inchoate cottages and schoolhouses" are the forests that will yield the lumber with which to build homes and schools. *Ingleside* is synonymous with *fireside*. (*Ingle*, which derives from Scottish Gaelic, means "fireplace or hearth.")

tle foothold. All these lands are worthless.[21] In passing from the Mexican to the British line, where conditions of altitude are the same, the grasses steadily improve, and those of the northern half are comparatively rich. But even here there are waste places, for lava-fields abound that are virtually desert. And there are "bad lands" that yield little vegetation. These lands are hills of clay and sand that are washed by the storms and baked by the sun. When the rains come the hillsides are sloughs, and when the winds come the dried surfaces crack and crumble. Then there are cañon lands that are carved by many winding, branching gorges, and thus are rendered worthless. Then there are alcove lands where every rill of the rainy season heads in a precipitous, rocky gulch. These are also barren. Then buttes are scattered over the mesas and plateaus—fragments of formations left by the destroying storms for their future employment. Then there are cinder-cones, naked and desolate. Often lines of cliff stretch athwart the country—the margins of mesas and plateaus. These cliffs are worthy of further mention. When the winds drift the clouds along the lowlands, such a cliff, a few hundred or a few thousand feet in height, obstructs their way. So the clouds rise and discharge their moisture, and floods are speedily born. In regions of cliff a large portion of precipitation is along these lines, and yet with this increased precipitation they are not favored with great vegetation, for the water glides away on the steep declivities, and a zone of lowlands near by receives them, and here the most valuable forests of piñon and cedar are often found. Then the mountains are not all grassy slopes, for they are often interrupted with rocks and ledges and cliffs that are naked.

Though the grasses of the pasturage lands of the West are nutritious, they are not abundant, as in the humid valleys of the East. Yet they have an important value. These grasses are easily destroyed by improvident pasturage, and they are then replaced by noxious weeds. To be utilized they must be carefully protected, and grazed only in proper seasons and within prescribed limits. But they cannot be inclosed by fences in small fields. Ten, twenty, fifty acres are necessary for the pasturage of a steer; so the

---

[21] Although some of these "worthless" lands are today national parks and recreation areas, others have become military reservations, atomic test sites, and dumps. Many have indeed been treated as worthless.

grasses can be utilized only in large bodies, and be fenced only by townships or tens of townships. Yet they must have protection or be ruined, and they should be preserved as one great resource of food for the people.[22] When the valleys below are irrigated, so that flocks and herds may be fed when the snows and frosts of winter come, the hills and mountains of the arid region will support great numbers of horses, cattle, and sheep.

The mountains of the far West are full of gold. Ores of the yellow metal are found in fissures that seam the rock, and fill spaces between barren formations, and lie in bodies where lavas have cooled in hill-bound basins. Then the whole mountain region has been plowed with glaciers and swept by storms or buried by river floods, and in these glacial gravels and storm gravels and river gravels the gold has been carried, and here the placer mines are found. In other hills and mountains there are stores of silver and copper, while lead and iron abound. Then asphalt, oil, and gas are found, and the hills are often filled with coal. With slight exception all of these minerals are found in lands which cannot be redeemed for agriculture. The coal lands are chiefly pasturage lands, and the gold and silver mines are under the forests. The coal and iron have been and can be discovered by science, but gold and silver are discovered by prospectors and revealed only by the pick and shovel. These mines of gold and silver furnish the basis of our monetary system, and are the source of vast wealth. During the last calendar year [1889] $32,816,500 in gold and $59,118,000 in silver were taken from these regions, and this supply is to be continued through an indefinite future.

---

[22] Abuse of the open range of the public domain was well advanced by 1890 but still had far to run. Although some public grazing lands received at least partial protection through withdrawal as Forest Reserves and, later, National Forests, the greater part of the public domain suffered a continuing tragedy of the commons for decades more. The open range system essentially punished stockmen if they rested their ranges or stocked them lightly: what they did not use, they would lose, for any available capacity might quickly be appropriated by another rancher. From Powell's time onward, Congress entertained dozens of bills to regulate grazing on the public domain but failed to enact any of them, mainly because of populist fear that powerful interests would gain control of grazing leases and squeeze out smaller operators, thereby also foreclosing the homesteading hopes of countless others who imagined moving to the promising West. It was not until Dust Bowl grime rained on the national capital that the 1934 Taylor Grazing Act was enacted, establishing a system of leases for grazing on the public domain. Powell had argued for a leasing system as early as 1878 in his *Report on the Lands of the Arid Region.*

When the waters are stored in the mountain lakes, and the canals are constructed to carry them to the lands below, a system of powers will be developed unparalleled in the history of the world. Here, then, factories can be established, and the rivers be made to do the work of fertilization, and the violence of mountain torrents can be transformed into electricity to illumine the villages, towns, and cities of all that land.

Such are the non-agricultural lands of the arid region. They are forest, pasturage, and mineral lands, on which great industries are in process of foundation. More than twenty years ago I entered the region for the purpose of studying its resources. The investigations then begun have been continued to the present time, and in them many of the great scientific men of America have been employed. In that early day gold and silver mining was the chief attraction, and there were inchoate cities and towns in many places. Agriculture and manufacturing were almost wholly neglected. Everywhere men were digging into the heart of the mountains for gold and silver, and armies of men were engaged in prospecting, lured, now here, now there, by rumors of great discoveries. These armies were composed of stalwart men, adventurous, brave, and skillful. Away in the wilderness, without capital, but endowed with brawn and brain, they established industries, organized institutions, and founded a civilization which must forever be the admiration of mankind. The physical conditions which exist in that land, and which inexorably control the operations of men, are such that the industries of the West are necessarily unlike those of the East, and their institutions must be adapted to their industrial wants. It is thus that a new phase of Aryan civilization is being developed in the western half of America.[23] On this subject I hope to be heard at another time.

---

[23] For Powell and for most of his contemporaries, *Aryan* was as much a cultural as a biological concept, which should not be confused with twentieth-century Aryanism and its virulent racism. Although Powell believed that western European cultural traditions were superior to others, this conviction did not lead him to conclude that other groups might justifiably be exploited or abused. His views on the relationships between humankind's cultural and ethnic divisions are expressed in Part VII.

---◇---

Institutions for the Arid Lands

---

*In this concluding article of the Century series, Powell outlines his vision for
a system of watershed commonwealths throughout the West.*

Industrial civilization in America began with the building of log cabins.
Where Piedmont plain merges into coastal plain, there rivers are transformed—dashing waters are changed to tidal waters and navigation heads,
and there "powers" are found.[1] Beside these transformed waters, in a narrow zone from north to south across the United States, the first real settlement of the country began by the building of log-cabin homes. This first
cabin zone ultimately became the site of the great cities of the East—
Boston, New York, Philadelphia, Baltimore, Washington, Richmond,
Charleston, and Augusta; and steadily, while the cities were growing, the
log-cabin zone moved westward until it reached the border of the Great
Plains, which it never crossed.

The arid region of the West was settled by gold and silver hunters aggregated in comparatively large bodies. Their shanties of logs, slabs, boards,
and adobes were speedily replaced by the more costly structures of towns
and cities, which suddenly sprang into existence where gold or silver was
found. These towns and cities were thus scattered promiscuously through

Century Magazine 40 (May 1890): 111–16.

[1] Powell is speaking of the fall line, which marks the upstream limit of conventional navigation and divides coastal plain from piedmont. By "powers" he means sites where
hydropower may be harnessed (i.e., for water-driven mills or, increasingly from the time of
this article onward, for generating electricity).

the mountain land. In them avenues of trees were planted and parks were laid out, and about them gardens, vineyards, and orchards were cultivated. From this horticulture sprang the agriculture of the region.

In the East the log cabin was the beginning of civilization; in the West, the miner's camp. In the East agriculture began with the settler's clearing; in the West, with the exploitation of wealthy men.[2] In the log-cabin years a poor man in Ohio might clear an acre at a time and extend his potato-patch, his cornfield, and his meadow from year to year, and do all with his own hands and energy, and thus hew his way from poverty to plenty. At the same time his wife could plant hollyhocks, sweet-williams, marigolds, and roses in boxed beds of earth around the cabin door. So field and garden were all within the compass of a poor man's means, his own love of industry, and his wife's love of beauty. In western Europe, where our civilization was born, a farmer might carry on his work in his own way, on his own soil or on the land of his feudal lord, and in the higher phases of this industry he could himself enjoy the products of his labor, subject only to taxes and rents. Out of this grew the modern agriculture with which we are so familiar in America, where the farmer owns his land, cultivates the soil with his own hands, and reaps the reward of his own toil, subject only to the conditions necessary to the welfare of the body politic.

The farming of the arid region cannot be carried on in this manner. Individual farmers with small holdings cannot sustain themselves as individual men; for the little farm is, perchance, dependent upon the waters of some great river that can be turned out and controlled from year to year only by the combined labor of many men. And in modern times great machinery is used, and dams, reservoirs, canals, and many minor hydraulic appliances are necessary. These cost large sums of money, and in their construction and maintenance many men are employed. In the practice of agriculture by irrigation in high antiquity, men were organized as communal bodies or as

---

[2] Powell passes over the agriculture of the Pueblos, the Yuman-speaking tribes of the lower Colorado, the O'odham in Arizona, and other Indian groups. He discusses the Hispanic settlers of the Southwest later in this article.

slaves to carry on such operations by united labor. Thus the means of obtaining subsistence were of such a character as to give excuse and cogent argument for the establishment of despotism. The soil could be cultivated, great nations could be sustained, only by the organization of large bodies of men working together on the great enterprises of irrigation under despotic rulers. But such a system cannot obtain in the United States, where the love of liberty is universal.

What, then, shall be the organization of this new industry of agriculture by irrigation? Shall the farmers labor for themselves and own the agricultural properties severally? or shall the farmers be a few capitalists, employing labor on a large scale, as is done in the great mines and manufactories of the United States? The history of two decades of this industry exhibits this fact: that in part the irrigated lands are owned and cultivated by men having small holdings, but in larger part they are held in great tracts by capitalists, and the tendency to this is on the increase. When the springs and creeks are utilized small holdings are developed, but when the rivers are taken out upon the lands great holdings are acquired; and thus the farming industries of the West are falling into the hands of a wealthy few.

Various conditions have led to this. In some portions of the arid region, especially in California, the Spanish land grants were utilized for the purpose of aggregating large tracts for wholesale farming. Sometimes the lands granted to railroads were utilized for the same purpose. Then, to promote the irrigation of this desert land, an act was passed by Congress giving a section of land for a small price to any man who would irrigate it.[3] Still other lands were acquired under the Homestead Act, the Preëmption Act, and the Timber-Culture Act.[4] Through these privileges title could be secured

---

[3] The 1877 Desert Land Act provided for conveyance of title to 640 acres (1 square mile) if the person or entity filing on the land provided evidence of irrigating it and paid aggregate fees of $1.25 per acre.

[4] The 1841 Preemption Act allowed squatters on the public domain to buy the land they occupied for $1.25 per acre, 160 acres per family. Under the 1862 Homestead Act settlers might obtain title to 160 acres of public land simply by occupying and "improving" it for a period of 5 years. In grudging but inadequate recognition of the insufficiency of the original homestead unit, Congress revised the act in 1909 to provide for 320-acre filings and in 1916

to two square miles of land by one individual. Companies wishing to engage in irrigation followed, in the main, one of two plans: they either bought the lands and irrigated their own tracts, or they constructed irrigating works and supplied water to the farmers. Through the one system land monopoly is developed; through the other, water monopoly.

Such has been the general course of the development of irrigation. But there are three notable exceptions. The people of the Southwest came originally, by the way of Mexico, from Spain, where irrigation and the institutions necessary for its control had been developing from high antiquity, and these people well understood that their institutions must be adapted to their industries, and so they organized their settlements as pueblos, or "irrigating municipalities," by which the lands were held in severalty[5] while the tenure of waters and works was communal or municipal. The Mormons, settling in Utah, borrowed the Mexican system. The lands in small tracts were held in severalty by the people, but the waters were controlled by bishops of the Church, who among the "Latter-Day Saints" are priests of the "Order of Aaron" and have secular functions. In southern California, also, many colonies were planted in which the lands were held in severalty and in small parcels.[6] Gradually in these communities the waters are passing from the control of irrigating corporations into the control of the municipalities which the colonies have formed. Besides these three great excep-

---

[4](*continued*) to permit 640-acre filings. The 1873 Timber Culture Act, which was intended to encourage settlement of the treeless Great Plains, provided settlers with an additional 160 acres if they planted trees on at least 40 acres (later reduced to 10 acres). Implicit in the Timber Culture Act was the belief, which Powell so often argued against, that widespread tree planting would substantially modify the climate of arid lands. In 1890 these acts would have provided for cumulative acquisition of 1.75 sections, not the 2 square miles mentioned by Powell. Powell may also have been thinking of the Timber and Stone Act of 1878, which allowed settlers in certain territories to purchase 160 acres of timber land for $2.50 per acre. In 1892 the act was extended to the public domain of all states and territories.

[5] "In severalty" means divided among multiple owners, each of whom has separate and private title to his or her individual ownership.

[6] For example, Etiwanda and Ontario in the Los Angeles Basin, which the entrepreneur George M. Chaffey established beginning in 1882. See Alexander, *The Life of George Chaffey*.

tions there are some minor ones, which need not here be recounted. In general, farming by irrigation has been developed as wholesale farming in large tracts or as wholesale irrigating by large companies. Some of these water companies are foreign, others are capitalists of the Eastern cities, while a few are composed of capitalists of the West.

Where agriculture is dependent upon an artificial supply of water, and where there is more land than can be served by the water, values inhere in water, not in land; the land without the water is without value. A stream may be competent to irrigate 100,000 acres of land, and there may be 500,000 acres of land to which it is possible to carry the water. If one man holds that water he practically owns that land; whatever value is given to any portion of it is derived from the water owned by the one person. In the far West a man may turn a spring or a brook upon a little valley stretch and make him a home with his own resources, or a few neighbors may unite to turn a small creek from its natural channel and gradually make a cluster of farms. This has been done, and the available springs and brooks are almost exhausted. But the chief resource of irrigation is from the use of the rivers and from the storage of waters which run to waste during a greater part of the year; for the season of irrigation is short, and during most of the months the waters are lost unless held in reservoirs. In the development of these water companies there has been much conflict.[7] In the main improvident franchises have been granted, and when found onerous the people have impaired or more or less destroyed them by unfriendly legislation and administration. The whole subject, however, is in its infancy, and the laws of the Western territory are inadequate to give security to capital invested in irrigating works on the one hand and protection to the farmer from extortion on the other. For this reason the tendency is to organize land companies. At present there is a large class of promoters who obtain options on lands and make contracts to supply water, and then enlist capital in the East and in Europe and organize and control construction companies, which, sometimes at least, make large profits. There seems to be little difficulty in interesting capitalists in these enterprises. The great

---

[7] For example, see Pisani, *To Reclaim a Divided West*, Chapter 4.

increase in value given to land through its redemption by irrigation makes such investments exceedingly attractive. But at present investors and farmers are alike badly protected, and the lands and waters are falling into the hands of "middlemen." If the last few years' experience throws any light upon the future the people of the West are entering upon an era of unparalleled speculation, which will result in the aggregation of the lands and waters in the hands of a comparatively few persons. Let us hope that there is wisdom enough in the statesmen of America to avert the impending evil.

Whence, then, shall the capital come? and how shall the labor be organized by which these 100,000,000 acres of land are to be redeemed? This is the problem that to-day confronts our statesmen and financiers. Capital must come, for the work is demanded and will pay. Let us look at the statistics of this subject in round numbers, and always quite within probable limits. Let us speak of 100,000,000 acres of land to be redeemed by the use of rivers and reservoirs.[8] This will cost about ten dollars per acre, or $1,000,000,000. In the near future a demand for this amount will be made, and it will be forthcoming beyond a peradventure. The experience obtained by the redemption of 6,000,000 acres of land, already under cultivation, abundantly warrants the statement that an average of fifty dollars per acre is a small estimate to be placed upon the value of the lands yet to be redeemed as they come to be used. Thus there is a prize to be secured of $5,000,000,000 by the investment of $1,000,000,000. Such vast undertakings will not be overlooked by the enterprising men of America.

In a former article on "The Irrigable Lands of the Arid Region," in *The Century Magazine* for March, 1890, it was explained that the waters of the arid lands flowing in the great rivers must somehow be divided among the States, and that in two cases important international problems are involved. It was also shown that contests are arising between different districts of the

---

[8]Although the most fervid proponents of irrigation believed Powell's estimate scandalously low, it was in fact too high. Less than 40 million acres are irrigated in the Arid Region today. See Selection 12, note 2. Powell's talk of irrigating 100 million acres of western land, meanwhile, alarmed eastern farmers and their political representatives, as it was equal to half of the nation's farmland then in production (Pisani, *To Reclaim a Divided West*, 164).

same State. But the waters must be still further subdivided in order that they may be distributed to individual owners. How can this be done? Lands can be staked out, corner-posts can be established, dividing lines can be run, and titles to tracts in terms of metes and bounds can be recorded. But who can establish the corner-posts of flowing waters? When the waters are gathered into streams they rush on to the desert sands or to the sea; and how shall we describe the metes and bounds of a wave? The farmer may brand his horses, but who can brand the clouds or put a mark of ownership on the current of a river? The waters of to-day have values and must be divided; the waters of the morrow have values, and the waters of all coming time, and these values must be distributed among the people. How shall it be done? It is proposed to present a plan for the solution of these problems, and others connected therewith, in an outline of institutions necessary for the arid lands. Some of these problems have been discussed in former articles, and it may be well to summarize them all once more, as follows:

First. The capital to redeem by irrigation 100,000,000 acres of land is to be obtained, and $1,000,000,000 is necessary.

Second. The lands are to be distributed to the people, and as yet we have no proper system of land laws by which it can be done.

Third. The waters must be divided among the States, and as yet there is no law for it, and the States are now in conflict.

Fourth. The waters are to be divided among the People, so that each man may have the amount necessary to fertilize his farm, each hamlet, town, and city the amount necessary for domestic purposes, and that every thirsty garden may quaff from the crystal waters that come from the mountains.

Fifth. The great forests that clothe the hills, plateaus, and mountains with verdure must be saved from devastation by fire and preserved for the use of man, that the sources of water may be protected, that farms may be fenced and homes built, and that all this wealth of forest may be distributed among the people.

Sixth. The grasses that are to feed the flocks and herds must be protected and utilized.

Seventh. The great mineral deposits—the fuel of the future, the iron for the railroads, and the gold and silver for our money—must be kept ready to the hand of industry and the brain of enterprise.

Eighth. The powers of the factories of that great land are to be created and utilized, that the hum of busy machinery may echo among the mountains—the symphonic music of industry.

A thousand millions of money must be used; who shall furnish it? Great and many industries are to be established; who shall control them? Millions of men are to labor; who shall employ them? This is a great nation, the Government is powerful; shall it engage in this work? So dreamers may dream, and so ambition may dictate, but in the name of the men who labor I demand that the laborers shall employ themselves; that the enterprise shall be controlled by the men who have the genius to organize, and whose homes are in the lands developed, and that the money shall be furnished by the people; and I say to the Government: Hands off! Furnish the people with institutions of justice, and let them do the work for themselves. The solution to be propounded, then, is one of institutions to be organized for the establishment of justice, not of appropriations to be made and offices created by the Government.

In a group of mountains a small river has its source. A dozen or a score of creeks unite to form the trunk. The creeks higher up divide into brooks. All these streams combined form the drainage system of a hydrographic basin, a unit of country well defined in nature, for it is bounded above and on each side by heights of land that rise as crests to part the waters. Thus hydraulic basin is segregated from hydraulic basin by nature herself, and the landmarks are practically perpetual. In such a basin of the arid region the irrigable lands lie below; not chiefly by the river's side, but on the mesas and low plains that stretch back on each side. Above these lands the pasturage hills and mountains stand, and there the forests and sources of water supply are found. Such a district of country is a commonwealth by itself. The people who live therein are interdependent in all their industries. Every man is interested in the conservation and management of the water supply, for all the waters are needed within the district. The men who control the farming below must also control the upper region where the waters are gathered from the heaven and stored in the reservoirs. Every farm and garden in the valley below is dependent upon each fountain above.

All of the lands that lie within the basin above the farming districts are the catchment areas for all the waters poured upon the fields below. The waters that control these works all constitute one system, are dependent one upon another, and are independent of all other systems. Not a spring or a creek can be touched without affecting the interests of every man who cultivates the soil in the region. All the waters are common property until they reach the main canal, where they are to be distributed among the people. How these waters are to be caught and the common source of wealth utilized by the individual settlers interested therein is a problem for the men of the district to solve, and for them alone.

But these same people are interested in the forests that crown the heights of the hydrographic basin. If they permit the forests to be destroyed, the source of their water supply is injured and the timber values are wiped out. If the forests are to be guarded, the people directly interested should perform the task. An army of aliens set to watch the forests would need another army of aliens to watch them, and a forestry organization under the hands of the General Government would become a hot-bed of corruption; for it would be impossible to fix responsibility and difficult to secure integrity of administration, because ill-defined values in great quantities are involved.[9]

Then the pasturage is to be protected. The men who protect these lands for the water they supply to agriculture can best protect the grasses for the summer pasturage of the cattle and horses and sheep that are to be fed on their farms during the months of winter. Again, the men who create water powers by constructing dams and digging canals should be permitted to

---

[9] Less than a year after this article appeared, president Benjamin Harrison ordered the withdrawal from the public domain of the first federal forest reserves. Before his term was over, Harrison established 15 reserves totaling 16 million acres. Under his successors the system grew to 41 reserves with 46.4 million acres by 1901, when Theodore Roosevelt became president. Roosevelt expanded the reserved area to 150.8 million acres by 1907, and to manage it he created a permanent federal bureaucracy, the U.S. Forest Service, in 1905, 3 years after Powell's death.

utilize these powers for themselves, or to use the income from these powers which they themselves create, for the purpose of constructing and maintaining the works necessary to their agriculture.

Thus it is that there is a body of interdependent and unified interests and values, all collected in one hydrographic basin, and all segregated by well-defined boundary lines from the rest of the world. The people in such a district have common interests, common rights, and common duties, and must necessarily work together for common purposes. Let such a people organize, under national and State laws, a great irrigation district, including an entire hydrographic basin, and let them make their own laws for the division of the waters, for the protection and use of the forests, for the protection of the pasturage on the hills, and for the use of the powers. This, then, is the proposition I make: that the entire arid region be organized into natural hydrographic districts, each one to be a commonwealth within itself for the purpose of controlling and using the great values which have been pointed out. There are some great rivers where the larger trunks would have to be divided into two or more districts, but the majority would be of the character described. Each such community should possess its own irrigation works; it would have to erect diverting dams, dig canals, and construct reservoirs; and such works would have to be maintained from year to year. The plan is to establish local self-government by hydrographic basins.

Let us consider next the part which should be taken by the local governments, the State governments, and the General Government in the establishment and maintenance of these institutions. Let there be established in each district a court to adjudicate questions of water rights, timber rights, pasturage rights, and power rights, in compliance with the special laws of the community and the more general laws of the State and the nation. Let there be appeal from these lower courts to the higher courts. Let the people of the district provide their own officers for the management and control of the waters, for the protection and utilization of the forests, for the protection and management of the pasturage, and for the use of the powers; and with district courts, water masters, foresters, and herders they would be equipped with the local officers necessary for the protection of their own property and the maintenance of individual rights. The interests are theirs, the rights are theirs, the duties are theirs; let them control their own actions. To some extent this can be accomplished by coöperative labor; but ultimately and gradually great capital must be employed in each district. Let them obtain this capital by their own enterprise as a community. Constituting a body corporate, they can tax themselves and they can borrow moneys. They have a basis of land titles, water rights, pasturage rights, forest rights, and power rights; all of these will furnish ample security for the necessary investments; and these district communities, having it in their power to obtain a vast increment by the development of the lands, and to distribute it among the people in severalty, will speedily understand how to attract capital by learning that honesty is the best policy.

Each State should provide courts for the adjudication of litigation between people of different districts, and courts of appeal from the irrigation district courts. It should also establish a general inspection system, and provide that the irrigation reservoirs shall not be constructed in such a manner as to menace the people below and place them in peril from floods.[10] And finally, it should provide general statutes regulating water rights.

But the General Government must bear its part in the establishment of the institutions for the arid region. It is now the owner of most of the lands, and it must provide for the distribution of these lands to the people in part, and in part it must retain possession of them and hold them in trust for the districts. It must also divide the waters of the great rivers among the States. All this can be accomplished in the following manner. Let the General Government make a survey of the lands, segregating and designating the irrigable lands, the timber lands, the pasturage lands, and the mining lands; let the General Government retain possession of all except the irrigable lands,

---

[10] Powell discusses the disastrous Johnstown Flood of May 1889 in "The Lesson of Conemaugh" (see Selection 10).

but give these to the people in severalty as homesteads.[11] Then let the General Government declare and provide by statute that the people of each district may control and use the timber, the pasturage, and the water powers, under specific laws enacted by themselves and by the States to which they belong. Then let the General Government further declare and establish by statute how the waters are to be divided among the districts and used on the lands segregated as irrigable lands, and then provide that the waters of each district may be distributed among the people by the authorities of each district under State and national laws. By these means the water would be relegated to the several districts in proper manner, interstate problems would be solved, and the national courts could settle all interstate litigation.

But the mining industries of the country must be considered. Undeveloped mining lands should remain in the possession of the General Government, and titles thereto should pass to individuals, under provisions of statutes already existing, only where such lands are obtained by actual occupation and development, and then in quantities sufficient for mining purposes only. Then mining regions must have mining towns. For these the townsite laws already enacted provide ample resource.

It is thus proposed to divide responsibility for these institutions between the General Government, the State governments, and the local governments. Having done this, it is proposed to allow the people to regulate their own affairs in their own way—borrow money, levy taxes, issue bonds, as they themselves shall determine; construct reservoirs, dig canals, when and

[11] This is one of Powell's most interesting ideas and constitutes a significant departure from his recommendations in the *Report on the Lands of the Arid Region*. Permanent federal ownership of grazing and timber lands would provide a check against alienation of those lands from the watershed community and prevent, for instance, outright sale of land to a timber company. Perhaps significantly, federal trusteeship would also have provided a toehold for future assertion of federal interests in management (e.g., to establish minimum standards for reforestation), although it is not known whether Powell had this in mind. In making this recommendation, Powell may have drawn upon Spanish and Mexican community land grants of New Mexico as a model: Irrigable lands were distributed in severalty, and the rest of the grant was a commons open to the use of all. Under Spanish and Mexican law, the land grant commons was indivisible and inalienable from the communities to which it belonged.

how they please; make their own laws and choose their own officers; protect their own forests, utilize their own pasturage, and do as they please with their own powers; and to say to them that "with wisdom you may prosper, but with folly you must fail."

It should be remembered that the far West is no longer an uninhabited region. Towns and cities are planted on the mountain sides, and stupendous mining enterprises are in operation. On the streams saw-mills have been erected, and the woodsman's ax echoes through every forest. In many a valley and by many a stream may be found a field, a vineyard, an orchard, and a garden; and the hills are covered with flocks and herds. In almost every hydrographic basin there is already found a population sufficient for the organization of the necessary irrigation districts. The people are intelligent, industrious, enterprising, and wide awake to their interests. Their hearts beat high with hope, and their aspirations are for industrial empire. On this round globe and in all the centuries of human history there has never before been such a people. Their love of liberty is unbounded, their obedience to law unparalleled, and their reverence for justice profound; every man is a freeman king with power to rule himself, and they may be trusted with their own interests.

Many of the great industrial undertakings of mankind require organized labor, and this demand grows with the development of inventions and the use of machinery. The transfer of toil from the muscles of men to the sinews of nature has a double result—social solidarity is increased, and mind is developed. In the past, civilization has combined the labor of men through the agency of despotism; and this was possible when the chief powers were muscular. But when the physical powers of nature are employed and human powers engaged in their control men cannot be enslaved; they assert their liberty and despotism falls. Under free governments the tendency is to transfer power from hereditary and chosen rulers to money kings, as the integration of society in industrial operations is accomplished through the agency of capital. This organization of physical power with human industry for great ends by the employment of capital is accomplished by instituting corporations. Corporations furnish money and machinery, and employ men organized under superintendents, to accomplish the works necessary to our modern civilization. Gradually society is being organized into a congeries of such corporations to control the leading industries of the land.[12]

Hitherto agriculture in this country has not come under the domination of these modern rulers. Throughout all the humid regions the farmer is an independent man, but in the arid regions corporations have sought to take control of agriculture. This is rendered possible by the physical conditions under which the industry is carried on. Sometimes the corporations have attempted to own the lands and the water, and to construct the great works and operate them as part and parcel of wholesale farming. In other cases the corporations have sought to construct the works and sell the water to individual farmers with small holdings. By neither of these methods has more than partial success been achieved. There is a sentiment in the land that the farmer must be free, that the laborer in the field should be the owner of the field. Hence by unfriendly legislation and by judicial decision—which ultimately reflect the sentiment of the people—these farming corporations and water corporations of the West have often failed to secure brilliant financial results, and many have been almost destroyed. Thus there is a war in the West between capital and labor—a bitter, relentless war, disastrous to both parties. The effort has been made to present a plan by which the agriculture of the arid lands may be held as a vast field of exploitation for individual farmers who cultivate the soil with their own hands; and at the same time and by the same institutions to open to capital a field for safe investment and remunerative return, and yet to secure to the toiling farmers the natural increment of profit which comes from the land with the progress of industrial civilization.

The great enterprises of mining, manufacturing, transporting, exchanging, and financiering in which the business kings of America are engaged

---

[12] Powell is evoking the monopolistic trusts that emerged in the latter decades of the nineteenth century, and against which Progressives such as Teddy Roosevelt campaigned. In the continuation of this passage he eloquently expresses his populist values and his belief that the government, through the institution of appropriate laws, is responsible to protect ordinary men and women from abuse by corporations and to reserve for them "the natural increment of profit" that comes from their labor.

challenge admiration, and I rejoice at their prosperity and am glad that blessings thus shower upon the people; but the brilliancy of great industrial operations does not daze my vision. I love the cradle more than the bank counter. The cottage home is more beautiful to me than the palace. I believe that the school-house is primal, the university secondary; and I believe that the justice's court in the hamlet is the only permanent foundation for the Supreme Court at the capital. Such are the interests which I advocate. Without occult powers of prophecy, the man of common sense sees a wonderful future for this land. Hard is the heart, dull is the mind, and weak is the will of the man who does not strive to secure wise institutions for the developing world of America.

The lofty peaks of the arid land are silvered with eternal rime; the slopes of the mountains and the great plateaus are covered with forest groves; the hills billow in beauty, the valleys are parks of delight, and the deep cañons thrill with the music of laughing waters. Over them all a clear sky is spread, through which the light of heaven freely shines. Clouds rarely mask the skies, but come at times like hosts of winged beauty floating past, as they change from gray to gold, to crimson, and to gorgeous purple. The soul must worship these glories, yet with the old Scotch poet[13] I can exclaim:

It's rare to see the Morning bleeze,

Like a bonfire frae the sea;

It's fair to see the burnie kiss

The lip o' the flow'ry lea;

An' fine it is on green hillside,

Where hums the bonnie bee,

But rarer, fairer, finer far

Is the ingleside for me.

---

[13] Searches of the work of Sir Walter Scott (1771–1832) and Robert Burns (1759–1796) have failed to find the source of these lines. Powell particularly favored Scott. On calm stretches during the second Colorado River exploration, he read to the men from *The Lady of the Lake.* In Scottish dialect, *bleeze = blaze, frae = from, burnie = brook,* and *ingleside = hearthside.*

PART VII

A Philosopher for
Humankind

F ROM THE MID-1880s onward, John Wesley Powell devoted the greater part of his creative energy to anthropological and philosophical questions. Even as the Geological Survey spawned the Irrigation Survey and as the latter blazed across the heavens of controversy, burning out a brief but intense life, Powell reflected and wrote on problems of increasing abstraction: How does culture evolve? What are the essential denominators of human experience? Are the nature of consciousness and the consciousness of nature the same?

The essays and one book that resulted from these ruminations are astonishing in their abundance and expansiveness, impressive in their length, and not a little soporific in their effect. Few among the many hundreds of pages Powell published in this area claim our attention today. Part of the reason is that the study of humankind and its ideas has changed radically since Powell's day. Like others of his century and the century before, Powell strove to integrate all knowledge into a single system, all sciences into one science. The twentieth century's explosion of knowledge soon rendered such efforts futile. By contrast, Powell's writing on the American West remains relevant because, although the circumstances of settlement have changed, the underlying problems of political organization and geographic adaptation have not. Americans are still settling both the land and the issues that attend their use of it.

Matters of style also may repel us from Powell's "scientific" writing, as he would have called his anthropological and philosophical papers. When appealing to the public or addressing essentially political issues, his language is reasonably, often admirably economical. When directing his thoughts to a scientific audience, however, Powell abandons concision for thoroughness. He unfolds his argument patiently, using no rhetorical shorthand lest he be misunderstood. The reader may, for instance, anticipate that what Powell has said about the plant world will apply in analog to animals, but Powell does not anticipate that anticipation; he explains the matter anew in the same terms, changing only a word or two. So exhaustive an approach becomes, for many readers, merely exhausting.

This is unfortunate because much of value and interest abides in Powell's anthropological and philosophical papers. He may have seen them as mainly scientific, as explorations of the phenomena of life, but from our vantage more than a century later, we can appreciate them better as professions of his most deeply held beliefs and as vigorous contributions to the important debates of his day, which are the ancestors of controversies that occupy us today. Here we meet the values of John Wesley Powell, as clearly and as directly as we meet them anywhere.

Underlying Powell's drive to classify and organize the things he studied—landforms, watersheds, Indian languages, and more—was his belief in the essential unity of all things. Philosophically, he was a monist. Science may have been eroding the old certainties of Mosaic Christianity, but he, like the other major thinkers of his day, still strove to place all knowledge in as integrated a framework as the divinity of the testaments had formerly provided. He proceeded from the conviction that there must be one matter, one species or family of forces shaping the diversity of the world's manifestations, one goal toward which the movement of creation, and especially human development, was directed.

Culture therefore was a singular, not a plural, phenomenon. One of the thinkers on whom Powell drew most heavily was Lewis Henry Morgan (1818–1881), a founder of modern anthropology and, like Powell, a native son of western New York. After intensive study of the Iroquois tribes of his home region, Morgan enlarged the scope of his field work to include the tribes o the Missouri River. He also solicited information about Native Americans from military officers and travelers to lands farther afield (a strategy Powell later institutionalized in the Bureau of American Ethnology). Through these direct and indirect means, Morgan strove to embrace in his researches native groups from almost every corner of the continent. He paid particular attention to kinship patterns: how families and clans were defined and how authorities of various kinds were distributed among their members. By analyzing and categorizing these patterns, Morgan provided the nascent science of anthropology with one of its most powerful classificatory tools. By further organizing these patterns into a single, linear theory of development, he also, it seemed to many, discovered a Rosetta stone for deciphering the evolution of human society. In 1877 he published *Ancient Society,* in which he posited how that evolution proceeded: It began in savagery, and then with the acquisition of certain skills and knowledge, it rose into barbarism, from which, with the blessing of further advances in technology and understanding, civilization ultimately blossomed. This hypothetically universal sequence of stages—from savagery to barbarism to civilization—became known as anthropology's classical evolutionary theory, and it dominated anthropological thinking for the decades of Powell's mature professional life.

Powell had heard of Morgan's formulation even before *Ancient Society* was published and began a long and admiring correspondence with him. When the book came out, Powell devoured it, reading it closely several times. Within a matter of months, he reported back to Morgan that the ethnographic information he had collected on the Colorado Plateau fit Morgan's framework handsomely. Here at last was a unifying theory around which to organize an abundance of information that otherwise seemed chaotically heterogeneous. From the appearance of *Ancient Society* onward, Powell became one of Morgan's staunchest allies, not only accepting the classical formulation of social evolution but bending his energy to detailing and fleshing it out.[1]

---

[1] Another ardent admirer of Lewis Henry Morgan was Karl Marx, who developed a theory of human social evolution that was similarly deterministic. Marx found especially interesting Morgan's assertion that bourgeois conventions of property ownership (in civilization) had been preceded by communal forms (in barbarism). Because Morgan was a Republican, a Presbyterian, and an anti-Socialist, Marx held his testimony to be unimpeachable.

The result, viewed more than a century later, is like a complex archaeo-logical site where evidence of one occupation is layered on another. Mor-gan's theory was built on intellectual terrain previously occupied by less enduring and more fanciful formulations, including, for instance, the notion that primitive people were degenerated branches of the same human line that had blossomed into European civilization and that a process of degeneration thus accounted for human diversity. Morgan's work swept such nonsense away and laid the foundation for a credible science of humankind. Efforts by Powell and others to elaborate classical evolutionary theory lent strength to Morgan's structure by helping to develop the skills of modern anthropology. Their attention to an ever-widening range of data, the discipline with which they organized their data, and their gradual, slow-growing sensitivity to ethnocentrism provided essential elements of mod-ern anthropological methodology. But their immediate successors—the anthropologists of the first half of the twentieth century—tore down or abandoned most of the theoretical structure they erected.

Morgan and his followers built their theory on an assumption of evolu-tionary unity: that all human societies developed along parallel paths in the same direction, and that because of favorable material conditions, some—such as European civilization in general and Anglo-American industrial society in particular—developed faster and attained a position superior to that of the others. Within scientific circles, this credo crumbled as the cen-tury turned. Its unitary view of cultural development could not bear the weight of new findings drawn from studies that were more genuinely com-parative than any Morgan or Powell had undertaken. Equally, the theory's implicit validation of industrial society as the desired endpoint of human development seemed less defensible as the excesses of industrialism mounted and the complexity of nonindustrial societies was better under-stood. The mantle of leadership in anthropology shifted to Franz Boas and his students, whose work embodied a powerful critique of imperialism and who showed that the paths and patterns of change in human society were more diverse than Morgan or Powell had imagined. They abandoned the-ory almost entirely, devoting themselves instead to "a raw, ethnographic 'natural history' approach to the study of primitive culture."[2] The unity for which nineteenth-century anthropologists had argued gave way to the relativistic idea that humankind had found multiple legitimate ways to skin the cat of life and answer questions of existence.

As a result, the first essay in this part, "From Barbarism to Civilization," seems more out of step with contemporary attitudes than the passage of time alone would suggest. It is a kind of bachelor uncle of modern anthropology: quirky and old-fashioned, memorable for the fellowship it once gave. But no lineal descendants perpetuate its name. Still, it bears a familial resemblance to much modern anthropological theory, controversial as that body of theory continues to be, for in the second half of the twentieth century evolutionary formulations about the development of human culture again drew adherents. These models of change were not orthogenetic, as was Morgan's. They accommodated far more complexity in the means and direction of social and cultural development, but while the new evolutionists rejected universal stages, they returned again to the conviction that the diversity of humankind might be understood in terms of repeated and identifiable patterns and processes. Morgan and Powell might not recognize the edifice that has since been erected (and continuously remodeled) on the foundation they labored to lay, but their work is in its base.

"From Barbarism to Civilization" remains interesting not merely as a datum in the history of anthropology. In its day the essay, as well as "Competition as a Factor in Human Evolution," which follows it, were part of a great debate, a contest for the moral center—call it the soul—of American society. And the echoes of that debate persist today.

No philosopher exerted more influence on American thought in the nineteenth century than the Englishman Herbert Spencer (1820–1903). Even Spencer's enemies had to admit that his philosophy, which found initial expression in *Social Statics* in 1851, was a marvel of synthesis and inclusion. The essence of Spencer's argument was that the laws governing the operation of the natural world governed also the workings of human soci-

---

[2] Service, "Cultural Evolution," 225.

ety. It was he, not Charles Darwin, who coined the phrase "survival of the fittest," and in that phrase one easily sees the germ of the virulent individualism that Spencer advised should be the model for human behavior. It was only natural, he said, that humans should vie with each other intensely, even brutally, for survival, and it was equally natural that the strongest should fare best. To interfere with that process was unnatural and counterproductive, he said, because interference would only prolong the unavoidable process of social adaptation that would lead ultimately to the highest possible development of civilization.

Consistent with this position, Spencer urged an extreme form of economic and social laissez-faire. In his view, government ought not attempt to regulate cutthroat business practices or to alleviate the suffering of the poor because the best long-term interest of society demanded that the struggle for existence proceed and the weak be sorted from the strong. (Voluntary charity, however, need not be neglected, because it had a salutary effect on those who provided it.) The moneyed classes on both sides of the Atlantic embraced such views during the latter half of the nineteenth century. They found in them a kind of congratulation, that to be wealthy meant, ipso facto, to be among the fittest. Here was a self-satisfying rationale for the continued and intense pursuit of self-interest. It was not only acceptable to be greedy for oneself and insensitive to the less fortunate; it was right to be so.

Spencer's philosophy was largely complete by the time Darwin's *On the Origin of Species* appeared in 1859, but Darwin's theory of natural selection, at least on its surface, seemed to ratify what Spencer had been saying. It was as though the men applied the same set of laws in different scientific arenas, Darwin in biology and Spencer in sociology, an emerging science that he did much to define. Critics of Spencer took advantage of this congruence. They branded him and his disciples *social Darwinists,* and the name stuck.[3]

---

[3] The seminal treatment of this subject is Richard Hofstadter's *Social Darwinism in American Thought.* Robert C. Bannister provides an insightful and important critique of Hofstadter in *Social Darwinism: Science and Myth in Anglo-American Social Thought,* and Eric Foner's 1992 introduction to the Beacon Press edition of Hofstadter's book places both approaches in perspective.

Anyone familiar with the *Report on the Lands of the Arid Region* or the work of the Geological Survey would understand that Powell found little to like in social Darwinism. He had argued all his professional life for government planning, government science, and the adoption of essentially communitarian forms of local government. He regarded the rampant individualism of his age, especially when played out in the context of antiquated land laws, as misguided and inappropriate. Powell accepted Darwin's evolutionary view and the principle of natural selection, but he rejected the arguments of Spencer and Spencer's most notable American disciple, William Graham Sumner (1849–1910). Society was not enslaved to a narrow reading of natural selection, he said, for selection also operated on ideas and technologies. This further evolutionary process, which Powell called human selection, enabled society to advance by cooperation as much as or more than by competition, and by creativity and intention as much as or more than by strife. As a result, "Hope for the future of society is the best-beloved daughter of Evolution."[4]

Throughout his professional life, Powell played mentor and friend to many talented individuals, and one in particular deserves mention here. In 1881, the Geological Survey hired Lester Frank Ward as assistant geologist and placed him in charge of paleobotany. Like Powell, Ward was a son of the midwestern frontier, a wounded combat veteran of the Civil War, and a self-taught master of multiple sciences. Before the decade was out, it was Ward, not Powell, who emerged as America's most persuasive critic of social Darwinism and as an ardent advocate of social activism. Among Ward's central messages was that the advancement of society depended on the knowledge and abilities of its people and that opportunities for education should therefore be made as widely available as possible. In this, he anticipated the later work of John Dewey. When Ward published his magnum opus *Dynamic Sociology* in 1883, he dedicated it to Powell. In 1898 Powell returned the favor and dedicated *Truth and Error*, his foray into a monist treatment of matter and consciousness, "to Lester F. Ward, philosopher and friend."

---

4 Powell, "Sketch of Lewis H. Morgan," 121.

Spencerism waned as the excesses of the Gilded Age grew more intolerable, but the debate over the superiority of individualist or communitarian values will never end. Societies will forever tinker with the balance they strike between the interests of the individual and the interests of the group, as each successive generation reexamines the rationale for the balance it inherits. What accounts for the differences between class and racial groups? Is it nature or nurture? The debate between Spencerians and advocates of human selection like Powell and Ward essentially parallels that between advocates of social laissez faire and social intervention in our own day.[5] For this reason, some of Powell's arguments in the following essays possess a surprisingly contemporary ring. The tone may be stiff, the point of view excessively masculine (Powell was a child of his time), and the treatment overlong (hence the omission of some repetitive sections, marked by ellipsis), but Powell's essential decency and democratic idealism nevertheless shine through.

Powell believed that the human spirit was capable of a kind of nobility that mere Spencerian self-interest was inadequate to explain: "The habit grows from generation to generation, until at last some men even forget that there is reward for service, and labor for their fellow-men because they love their fellow-men. They witness the disabilities of the maimed and they build them homes, for the sick they build hospitals, for the insane they erect asylums, for the blind and deaf they establish schools, for the ignorant they furnish instruction, for the wronged they enact laws and demand justice, for the enslaved they fight for liberty." Powell, an ardent abolitionist, knew something about fighting for the liberty of the enslaved, having learned a stern lesson at Shiloh. And more generally, he knew something about service to society, having devoted the work of his mature years toward that end. Few among us can help wishing that more men and women of Powell's stamp were serving society today.

---

[5] See Herrnstein and Murray, *The Bell Curve* (1994) and replies thereto, including Fischer et al., *Inequality by Design;* Devlin et al., *Intelligence, Genes, and Success;* and Kincheloe et al., *Measured Lies.*

From Barbarism to Civilization

Powell originally delivered this essay as his annual presidential address to the Anthropological Society of Washington on March 16, 1886.

The course of human events is not an eternal round. In the wisdom of the ancients there are many proverbs to the effect that that which is, has been before, and will be again. So far as human experience extends, unaided by reason, days and nights come and go, winter follows summer, and summer follows winter, and all the phenomena of nature seem to constitute an endless succession of recurrent events. There is a higher knowledge which observes a progress made only by steps so minute that it was left to modern science to discover them. In the history of humanity the changes which result in progress are more readily perceived, and the aphorism of the ancients that "There is nothing new under the sun" is but a proverb of ignorance. That which has been is not now, and that which is never will be again, in all the succession of phenomena with which anthropologists deal, for with apparent repetition there is always some observable change in the direction of progress.

Every child is born destitute of things possessed in manhood which distinguish him from the lower animals. Of all industries he is artless; of all institutions he is lawless; of all languages he is speechless; of all philoso-

The American Anthropologist 1/2 (April 1888): 97–123. The cited volume is in the first year of the "New Series" of *The American Anthropologist,* which in 1888 commenced serving as the combined organ of several learned societies including the American Anthropological Association and the Anthropological Society of Washington.

phies he is opinionless; of all reasoning he is thoughtless; but arts, institutions, languages, opinions and mentations he acquires as the years go by from childhood to manhood. In all these respects the newborn babe is hardly the peer of the new-born beast; but as the years pass, ever and ever he exhibits his superiority in all of the great classes of activities, until the distance by which he is separated from the brute is so great that his realm of existence is in another kingdom of nature. The activities that segregate mankind into a kingdom of his own are the humanities. The human race has been segregated from the tribes of beasts by the gradual acquisition of these humanities, namely: by the invention of arts; by the establishment of institutions; by the growth of languages; by the formation of opinions and by the evolution of reason. If this be true—and this is demonstrated by the science of anthropology—then the road by which man has traveled away from purely animal life must be very long; but this long way has its landmarks, so that it can be divided into parts. There are stages of human culture. The three grand stages have been denominated Savagery, Barbarism, and Civilization.

In the popular mind, and even in the conception of many scholars, there is no clearly perceived and well-defined distinction between these stages. There is a vague idea that the barbarian is somehow a lower being in the scale of existence than the civilized man, and that the savage is still lower—that the savage is savage, that the barbarian is barbaric. But no attempt has been made to set forth the characteristics of these stages of culture systematically in all the grand classes of activities, and thus they have never been thoroughly and satisfactorily defined. In my last annual address to this Society the endeavor was made to characterize savagery and barbarism, and to show the nature and origin of the change from the lower to the middle stage.[1] The endeavor will be made tonight to partially recharacterize barbarism in terms of the humanities, and then to show how all of these were changed and developed into something higher, that something higher being the proper attributes of civilization.

---

[1] Powell, "From Savagery to Barbarism."

Before proceeding to this task certain errors in the current literature of anthropology must be dispelled.

It must be remembered that in the attempt to define savagery, barbarism, and civilization, they are treated as stages of culture, not as characteristics of individuals. A stage of culture is represented by the aggregate of human activities, the humanities, extant among the people and during the time to which such characterization belongs.

It must be further remembered that a stage of society which extends perhaps over many lands and embraces many bodies politic and continues through many generations exhibits a vast variety of individual characters, and it would be an absurdity to claim that in each man in civilization we discover a full exemplification of all of the attributes of that civilization. No one man, even the greatest, has been equal to the whole of the men of his time, and there are always vast numbers who fall far short of acquiring the culture which properly characterizes their times. In every land and among every people there are some who are imbecile, depraved, or ignorant, and who thus utterly fail to exhibit the current culture. These are the degraded classes. But it should be noted that this degradation is not toward a more primitive culture. The vicious and ignorant in civilization do not lapse into barbarism by adopting the arts of barbarism, by establishing barbaric institutions, by returning to the use of barbaric languages, and by adopting the opinions of barbarism; but they fail in acquiring the culture of civilization by a failure in the acquisition of any culture. Retrogression in culture proper is rarely, perhaps never, exhibited on any large scale. The frequent failure of individuals to acquire the culture proper to their time and place in history has sometimes led to mistaken theories in regard to the general progress of culture, and in this manner the conclusion has been reached that there are progressive and retrogressive races and that culture itself waxes and wanes.

Civilized travelers among the lower races of mankind have often formed hasty judgments and have characterized peoples from the accidental observations of a day; but I think it may be safely asserted that no thorough study of any race or tribe has ever led to the discovery of an extended and continued loss of culture. The accounts of hasty travelers may be divided into two classes: In a general way, one set of writers have found among savage and barbaric peoples a state of affairs worthy only of execration, and all such peoples have thus been condemned as "devils;" another set of writers have discovered among such peoples only evidence of primitive innocence and the happiness of primitive simplicity, and such peoples have been pictured as "angels." But neither of these conclusions is reached by trained anthropologists whose studies of mankind are made by careful investigation. . . .

It is a subject of frequent observation and remark that ignorant people suppose that all languages other than their own are not real languages but only jargons, and that the tones of unfamiliar languages are not much better than brutish gruntings; in like manner the unfamiliar and misunderstood habits and customs of aliens and strangers are but absurdities to the ignorant, and to such persons all human activities other than their own are but the acts of fools.[2] Somewhat of the same nature are many current opinions of savage and barbaric life. In civilization tribal people have often been characterized with all the prejudice of ignorance. Now there is a cheap scholarship which goes far and wide to collect these prejudiced and ignorant statements and bases upon them a theory of savage culture. By these easy lessons it is discovered that savagery is a state of perpetual warfare; that the life of the savage is one of ceaseless bloodshed, that the men of this earliest stage of culture live but to kill and devour one another, and that infanticide is the common practice. Starting with man in this horrible estate these same scholars construct a theory of the evolution of mankind from savagery to civilization as the transition from militancy to industrial-

---

[2] Powell is describing what future generations would call *ethnocentrism*, the assumption that one's own cultural knowledge and habits are superior to others. Powell was exceptional in his time for his sensitivity to the presence of such a set of cultural blinders, yet at times he wore those blinders himself. His failure to attain perfection in this regard is perhaps less a testimony to his inconsistency than to the enduring difficulty of sensing the refraction of the cultural lens through which one sees the world.

ism. Such is the Spencerian philosophy of human development, and it has many adherents.[3]

Human industries, like other human activities, have had their course of evolution, and militancy itself has been developed from lowly beginnings to an advanced state of organization. The savage tribes of mankind carried on petty warfares with clubs, spears, and bows and arrows. But these wars interrupted their peaceful pursuits only at comparatively long intervals. The wars of barbaric tribes were on a larger scale and more destructive of life; but there were no great wars until wealth was accumulated and men were organized into nations. The great wars began with civilization, and have continued to the present time. Steadily armies have become larger, and more thoroughly organized as naval and land forces, and the land forces as infantry, artillery, and cavalry; and with the progress of civilization armies have been equipped with implements of warfare more and more destructive. Savage warfare compares with modern warfare as the bow and arrow compare with the Gatling gun; as the stone club compares with the sixteen-inch Krupp rifle. May be the nineteenth century has had greater armies than ever before existed, and these forces have been armed with more terrible implements of destruction than ever before known, and the sacrifice of human life in the nineteenth century has been greater perhaps than in any other such period of the history of the world. Warfare has had its course of evolution, as have all other human activities. That human progress has been from militancy to industrialism is an error so great that it must necessarily vitiate any system of sociology of theory of culture of which it forms a part.

The errors of the ignorant are often committed by inconsiderate travelers, who have reported that the tribes with which they met, now here, now

---

[3] Central to Herbert Spencer's philosophy was the notion that the evolution of civilization, subject to "natural law," led from the militaristic kingdoms of the past to an "equilibrated" social state that would be attained in the industrial future. In attaining this final stage, society would have made all possible adaptations and would therefore exist in a state of harmony, inwardly among its constituent parts and outwardly in relation to its environment. The belief that an ethic of "every man for himself" led inexorably to social perfection justified the extreme individualism of the Gilded Age and helped make Spencer the most influential thinker of the period.

there, were destitute of any real language; that they had a few grunts, exclamations, and jargon words, and eked these out by the use of gesture speech, and many able scholars have accepted these statements as facts. When a savage or barbaric tribe comes in contact with civilization there grows up between the two people a jargon of corrupted words derived from both languages. This jargon is always ephemeral and it rarely acquires the status of a language. Many travelers and scholars have mistaken such jargons for languages themselves and have inferred therefrom that tribal languages are exceedingly unstable. Again, the lower the grade of culture the smaller the number of people speaking one language. As we go back from civilization towards savagery, languages rapidly multiply, and this diversity of speech has strengthened some scholars in the notion that savage languages are rapidly changeable. Thus there is a tendency among philologists to depreciate savage tongues and to consider them as composed of few words and incapable of expressing any great body of thought and as rapidly changing from generation to generation. On the other hand, whenever a savage language is thoroughly studied, it is invariably found to have a copious vocabulary and to be highly organized by an indiscriminate variety of grammatic devices. When such languages are discovered, the difference between real savage languages and supposed savage languages is so great that at once retrogression is affirmed of them, the reasoning being something like this: "Savage languages proper are of this low class, the characteristics of which have been given us by these travelers; now, here is a language very much more highly developed: therefore, the people have been degraded from some civilized state."

In further support of this theory, the language itself is placed in a grade much higher than it deserves. A copious vocabulary is no evidence of high development. The law of gradation in this respect seems to be entirely misunderstood. The different thoughts possible even to savage minds are practically innumerable, and every language, even that of the savage, is capable of expressing all of the thoughts possible to the people who use the language. It is a characteristic of the languages of savages that many words are necessary to express their thoughts, while in civilized languages the same thoughts can be expressed with a smaller number of words.[4] Given a body of thought, then, that language is the most highly developed which uses the smallest number of words for its expression. This improvement in the language, by which the fewer words can be used for the greater expression, is accomplished by the organization of the language through the development of parts of speech and the integration of the sentence.

A language is high or low not by reason of the number of words which it contains but by reason of the degree of organization to which it has attained and the body of thought which it is competent to properly express. This may be made clear by an illustration from the written language of numbers. In a written language there might be a character for each of the numbers to a hundred, and to express multiples of such numbers repetitions of the characters might be necessary. The notation of such a language would thus have many characters, but it would not be highly developed. Again, the Roman system of numerals with which we are all acquainted has few characters but a very crude method of representing multiples of numbers by the use of these characters, and such a notation is of very low degree as compared with the Arabic system of notation, by which a few characters are used, the value of these depending upon their placement. In like manner in savage tongues there is a vast number of words which are exceedingly cumbrous when used in the expression of thought, from the fact that the parts of speech are not differentiated nor the sentence organized. Usually those who devote themselves thoroughly to the study of savage languages clearly understand their low character, while those who devote themselves to the study of the classical languages, having before them false models of excellence, seem always to exaggerate the

---

[4] Now Powell's ethnocentrism manifests, even as he refutes the ethnocentric errors of others. Could it be that the economy of expression in language is related not so much to high or low development, but to the degree to which the thoughts being expressed are native to the language? A description in English of how to build a metal lathe will likely use fewer words than an equally detailed description in Inuit, but if the item to be built is a kayak, Inuit will beat English handily.

value of the savage languages which have been thoroughly studied and to undervalue all other savage languages, holding them to be properly characterized by the ignorant travelers.

There is yet another class of errors to be noted. In the vast commingling of peoples through the enormous development of means of transportation in later civilization, everywhere savage and barbaric peoples are associated more or less with civilized men. In this association, the lower races always borrow something of arts, institutions, languages, and also of philosophic opinions—they borrow explanations of phenomena. Now, it is a curious fact that these borrowed opinions are often unrecognized as such by scholars, and hence savage and barbaric peoples are described as entertaining opinions far beyond the grade to which their indigenous culture would carry them. These savage peoples are again and again represented as believing in one God, as if in fact they were monotheists by autogenous[5] culture. They are also represented as believing in angels, as believing in heaven or a "happy hunting ground," and as believing many other things which pertain not to savagery and barbarism but to civilization.

To the metaphysician who juggles with formal logic, the light and the darkness can always be clearly distinguished as the light and the non-light or the darkness and the non-darkness. To the scientific man the absolute light and the absolute darkness are never found, but the phenomena of light and darkness cover infinite degrees of chiaroscuro, with absolute light on one hand and absolute darkness on the other, beyond the boundaries of observed phenomena and existent only in statement. To the scientific man it will not be necessary to explain that in defining stages of culture, types only are characterized, between which infinite gradations are found, but the metaphysician will doubtless come in with his formal logic and fail to discover absolute barbarism and absolute civilization. This exposition is for scientific men who deal with phenomena—let the jugglers juggle.

Having cleared the pathway through which we are to travel in the con-

---

5 Self-developed; not altered by outside influence.

sideration of this subject of the errors which cast deep shadows along the course, we can proceed to define "barbarism" and "civilization" and point out the course of cultural progress involved.

The Change in Arts

That which has elevated many of the tribes of mankind above savagery and into the stage which we call barbarism was the cultivation of the soil and the domestication of animals; and through these means their food-supply was greatly increased, and the more because the animals themselves were used as aids to agriculture. Yet further, horses and camels were used as means of transportation. Barbarians also have dwellings of wood and stone; often these dwellings are communal, and thus compound houses were constructed for clans and even for tribes. It is worthy of note, also, that in these houses there was a family hearth, for chimneys had been invented. Barbarians clothe themselves but slightly with furs, chiefly with textile fabrics, for they are skillful in spinning and weaving. They also have a good supply of culinary utensils, for they mould and burn clay and thus have abundant pottery. The barbarian not only has beasts of burden for means of transportation on land, but he traverses the rivers and meanders the shores of lakes and seas with boats propelled with oars. In this stage, the simple materials of nature on every hand are utilized—stone, copper, wood, shell, bone, and horn are fashioned into new shapes and often with great skill. But while they have tools they have no machinery; for example, there is no potter's wheel, no grist mill, and no saw mill. . . .[6]

---

[6] The sections omitted here include a large amount of material. "The Change in Arts" continues with a discussion of barbaric versus civilized tools, clothing, architecture, and pottery. "The Change in Institutions" surveys kinship and marriage, governance, warfare, tax collection, and the emergence of the nation state. "The Change in Language" emphasizes grammatical sophistication and the development of alphabets. "The Change in Opinions" explores the emergence of monotheism and increasing abstraction in "civilized" religion and philosophy. Powell concisely summarizes his generalizations about these subjects near the end of the essay.

The Change of Mentations

In barbarism men had learned the simple rules in arithmetic. The organization of the first great tax-gathering nation of which we have knowledge, on the banks of the Nile, ultimately led to the gathering of a vast population within the valley, so that it became necessary to utilize all the flood waters of the river in the fertilization of the soil to produce the food necessary to support the vast population gathered in the desert-bound valley. Then land mensuration was invented; and time-mensuration, as the people became dependent on the seasons; when geometry was invented mathematical reasoning was born.

When the great scholars of many nations had studied and compared many languages, many political systems and laws, and especially when they had studied and compared many systems of mythology and found that opinions were immeasurably diverse, they made the still deeper discovery that men had everywhere reasoned on false analogies; and in order that they might discover the truth itself, they sought to discover some method by which they could identify truth when found—some test of truth, some miraculous talisman that would glow when truth was spoken! In this search for the touchstone of truth, they said to themselves: "The human mind is primarily endowed with fundamental principles by which all opinions may be tested. Fundamental principles—major premises—are the glowing talismans of truth." It was introverted thought that led to this conclusion. With these primordial principles, which every thinker believed he discovered within himself, the machinery of formal logic was devised. This formal logic led to a new philosophy which never extended very far so as to lead great bodies of men into methods of interpreting nature. Still, it was a philosophy which gained credence among the few and by which some of the phenomena of nature received interpretation among these few. So strangely, indeed, early civilization developed mathematical reasoning by which the truth itself is reached and also metaphysical reasoning which leads only to the realm of phantasms.

The most important acquisition to intellectual activity ever gained by man is the power of inductive reasoning beyond the penetration of the senses and beyond sensuous conceptions and into a realm in which conclusions are reached which are apparently contradicted by the senses and by experience. Mr. Lester F. Ward, a vice president of this Society,[7] has with great acumen and skill shown how often the phenomena of nature are misleading as they are presented directly to the mind. To the senses the earth is standing still and the heavens are revolving about it. This is the direct teaching of our perceptions. But the reason is carried by many inductions to the conclusion that the seeming motions of the stars from east to west are indeed but measures of the motion of the earth on its axis from west to east. To all sensuous perceptions material things disappear and are annihilated—something becomes nothing. On the other hand, to the same sensuous perceptions creations appear; out of nothing something comes, and yet in spite of the constant averment of the senses the grand induction is reached that matter cannot be created or destroyed. That motion comes to an end is ever the experience of the senses. In fact it seems to be observed in every moment of wakefulness; and that it might be brought into existence from a state of rest seems to be a constant observation. Thus the senses attest to the belief that motion can be created or destroyed. Yet in spite of universal sensuous perceptions the grand induction has been reached that motion is persistent; and in general throughout the phenomena of nature, that which first appears to observation is but appearance; the verity is discovered only by profound investigation.

This power to reach inductive conclusions in opposition to current and constant sensuous perceptions is the greatest acquisition of civilized culture.

With the advent of civilization a new class of mental activities developed. The desire to know had existed throughout the whole course of culture, but in civilization this desire found new expression in scientific research. It

---

[7] By March 1886, when Powell gave this address to the Anthropological Society of Washington, Lester Ward (1841–1913), one of Powell's closest allies, had emerged as an effective critic of social Darwinism and a leading sociologist in his own right.

has already been noted that the arts of civilization made men travelers and thus made men scholars, for the travelers wherever they went found new arts, new institutions, new languages, and new opinions which challenged their attention and they were led into comparative studies. It is a curious fact that in the development of any new science the term comparative is usually taken as a part of its name, for it is at once seen that the new science arises out of the adoption of new lines of comparison. Thus quite early in civilization the science of comparative jurisprudence was developed. Out of this came what was called at first "the law of nations," which must not be confounded with international law, the law of nations being the formulated code or system of principles discovered to exist in the laws of all nations. In like manner the comparative study of languages led to early development of the science of philology; so at the very dawn of civilization many new sciences were born, men engaged in research for the sake of discovering new truths and in so doing sought for and collected new facts and made new and various comparisons, and it was in this manner that scientific research was instituted. Along the entire course of culture it had been well known that knowledge would be utilized to secure benefits; in civilization it was discovered that knowledge could be increased and increasing benefits would result therefrom. In the earlier stages of culture, invention led to discovery. The utilization of knowledge led to new knowledge; but in civilization research led the way and the increasing knowledge was utilized by invention to secure new benefits. Thus research increases science, and science is applied to useful purposes, and scientific research belongs to civilization only. Early invention grew out of the physical wants and desires of mankind and was directed towards their supply. In civilization invention was directed into new channels and man began to invent methods and instruments of research. Thus it was that the mental activities of man were greatly increased by civilization.

It is not a part of the present task to exhibit the course of intellectual progress through civilization; this would lead to the history of the invention of scientific instruments and appliances for investigation by which the human senses are extended into realms of perception at first unknown to

civilization; such, for example, as the invention of the telescope, the microscope, and the spectroscope, and the method by which the earth is weighed
with the pendulum, and by which, with measured angles and mathematical
formulae, the distances of the stars are learned. It would also lead us to a
consideration of the growth of inductive reasoning, by which facts are classified into groups and groups of groups are made until conclusions are
reached of such great proportions that only minds trained by the handling
of these inductions through all their stages are able to grasp them. It would
lead us to a consideration of the methods of reasoning by which the apparent dome of the heavens has been resolved into an infinitude of space filled
with an infinity of worlds. It would also lead us to a consideration of the
methods by which the mountains and the lakes and the seas and all material
objects have been resolved in knowledge into minute and almost infinitesi-
mal atoms and by which heat, light, and electricity have been discovered to
be but modes of motion. It would lead us also into consideration of the subject of the evolution of worlds and the evolution of life itself. All of this
belongs to the history of civilization.

Let us return then, to a brief characterization of savagery, barbarism, and
civilization; and in order that it may be laconic, all qualifications and provi-
sos must be neglected.

The age of savagery is the age of stone; the age of barbarism the age of
clay; the age of civilization the age of iron.

The savage propels his canoe with a paddle; the barbarian propels his
boat with oars; the civilized man navigates the sea with ships propelled by
sails.

In savagery, music is only rhythm; in barbarism it is rhythm and melody;
in civilization it is rhythm melody, and harmony.

The age of savagery is the age of kinship clan, when maternal kinship is
held most sacred; the age of barbarism is the age of kinship tribes, when
paternal kinship is held most sacred; the age of civilization is the age of
nations, when territorial boundaries are held most sacred.

In savagery, law is designed to secure peace; in barbarism, to secure
peace and authority; in civilization, to secure peace, authority and justice.

In savagery, law extends only to kindred; in barbarism, to kindred and retainers; in civilization, to all the people of the nation.

The age of savagery is the age of sentence words; the age of barbarism the age of phrase words; the age of civilization the age of idea words.

In savagery, picture-writings are used; in barbarism, hieroglyphs; in civilization, alphabets.

In savagery, there is no verb "to be"; in barbarism, there is no verb "to read"; in civilization, verbs are resolved into parts of speech.

In savagery, beast polytheism prevails; in barbarism, nature polytheism; in civilization, monotheism.

In savagery, a wolf is an oracular god; in barbarism, it is a howling beast; in civilization, it is a connecting link in systematic zoology.

In savagery, the powers of nature are feared as evil demons; in barbarism, the powers of nature are worshiped as gods; in civilization, the powers of nature are apprenticed servants.

In savagery, men can only count; in barbarism, they have arithmetic; in civilization, they understand geometry.

In savagery, vision is limited by opinion; in barbarism, vision is limited by horizon; in civilization, vision is limited by the powers of the telescope and microscope.

In savagery, reason is based on zoomorphic analogies; in barbarism, on anthropomorphic analogies; in civilization, on intrinsic homologies.

To those who have heard my addresses on this subject and by whom I have not failed to be understood it will appear that I have denied many of the fundamental propositions of that school of philosophy which extends the methods of biotic evolution to the realm of mankind.[8] I have affirmed that the man and the beast belong to different kingdoms of nature, and that the law of animal evolution is not the law of human progress. I have denied that man has progressed by the survival of the fittest in the struggle for existence, and I have affirmed that old philosophy that human progress is by human endeavor, exhibited in the effort to utilize the powers and materials

---

[8] Spencerism, or social Darwinism.

of nature by his inventions; in the effort to establish peace and justice; in the effort to express thought by the invention of language; in the effort to learn the truth by investigating the phenomena of the universe; and in every effort of intellectual activity. This same philosophy which affirms the futility of endeavor is the philosophy of "let alone."[9] It is the philosophy that asks the question of skepticism: "What do social classes owe to one another?"[10] It is the philosophy of the robber who fears to encounter the wronged owner; it is the philosophy of the murderer who asks the question of denial: "Am I my brother's keeper?" Metaphysics, the philosophy of Aristotle, was the cloud which hid the sun of truth from mankind through the middle—the dark ages. Should the philosophy of Spencer, which confounds man with the brute and denies the efficacy of human endeavor, become the philosophy of the twentieth century, it would cover civilization with a pall and culture would again stagnate. But science rends that pall, and mankind moves on to a higher destiny. Now, let me return to my theme:

The greatest intellectual discovery of savagery was the discovery of the difference between the animate and the inanimate, between the organic and inorganic, between the living world and the dead world; but the discovery having been made, the animals were deified and believed to be the authors and movers of the world of phenomena. The greatest intellectual achievement of barbarism was the discovery of the limited powers of animals; but the discovery having been made, the powers and wonders of nature were deified and given the forms of men. The greatest intellectual achievement of civilization was the discovery of the physical explanation of the powers and wonders of the universe, and the intellectual superiority of man, by which he becomes the master of those powers and the worker of wonders.

In savagery, the beasts are gods; in barbarism, the gods are men; in civilization, men are *as gods,* knowing good from evil.

---

[9]Laissez-faire.

[10] In 1883 William Graham Sumner published *What Social Classes Owe to Each Other,* a vigorous attack against calls for economic regulation and the alleviation of social problems. Sumner would have classed Powell's plan for the Arid Region among the attempts at social reform that he derogated as being misguided.

—◇—

Competition as a Factor in Human Evolution

———

This was Powell's address as the retiring president of the Anthropological Society of Washington, of which he was the leading founder and first president in 1879.

———

The world is endowed with abundant life. Living beings have wonderful powers of reproduction. If all the vegetation upon the surface of the earth were destroyed leaving but one young palm, one young oak, and one young pine, and if these were allowed to bear their fruits and every seed permitted to grow and reproduce its kind, in a succession of generations this palm, oak, and pine might live to see their progeny covering the whole earth— reforesting the whole surface of the land, and the younger palms and oaks and pines would stand so dense under the shadow of the taller forests that the world would be a jungle impenetrable to the larger beasts. Such are the powers of reproduction with which palms, oaks, and pines are endowed.

If a barrel of oysters were planted in an estuary of the sea and their progeny preserved in successive generations for ten years the oysterfield thus produced would supply a bounteous repast for every man, woman, and child on the face of the earth.

If in the economy of nature all of the worlds that are catalogued as stars were suddenly placed under the same conditions for the support of life as those existing on the earth, and then the earth were made the nursery of this universe, so that all of its redundant germs could be utilized on the stars,

The American Anthropologist 1/4 (October 1888): 297–323.

they would all in a few years be clothed with plants and the air and the land and the water of every star would teem with animal life.

It is beyond the capacity of the human mind to comprehend the powers of biotic reproduction. This marvelous fecundity, especially in the lower forms of life, has played an important part in the evolution of plants and animals, the nature of which must be understood. Few of these biotic germs reach adult life. For every successful passage through the term of life possible to the individual of each species there are a multitude of failures. The life of the few, and the very few, is secured by the martyrdom of the many, and the very many.

If many are called to life and few are chosen to live, how are the favored few selected? The answer to this question is the modern doctrine of evolution; it is "the survival of the fittest in the struggle for existence"; it is the philosophy that Darwin embodies in the phrase "natural selection." Nature gives more lives than she can support. There are more individuals requiring nourishment than there is food. Only those live that obtain sufficient nutriment, and only those live that find a habitat. Of the multitude of germs, some perish on the rocks, some languish in the darkness, some are drowned in the waters, and some are devoured by other living beings. The few live because they fall not on the rocks, but are implanted in the soils; because they are not buried in the darkness, but are bathed in the sunlight; because they are not overwhelmed by deep waters, but are nourished by gentle rains, or because they are not devoured by the hungry, but dwell among the living. The few live because they are the favorites of surrounding circumstances. In the more stately phrase of the philosophy of evolution, they are "adapted to the environment." Evolution, or progress in life, is accomplished among animals or plants by killing the weaker—the less favored—and by saving the stronger and more favored. Many must be killed because there are too many, and so the best only are preserved. Those a little above the average are saved, and this is called "natural selection." . . .

Competition among plants and animals is fierce, merciless, and deadly; out of competition fear and pain are born; out of competition come anger, and hatred, and ferocity. But it must not be forgotten that from this same competition there arise things most beautiful and lovely: the wing of the butterfly, the plumage of the bird, and the fur of the beast; the hum of the honeybee, the song of the nightingale, and the chatter of the squirrel. So good and evil dwell together.[1]

Having thus characterized the competition that obtains among plants and lower animals in their struggle for life, it is proposed to characterize the competition which exists in the higher civilization between man and man, and to show in what respect it may be like and in what respect it may be different from biotic, which exists in the lower orders of creation; and for this purpose the savage and barbaric tribes of men will be neglected; nor will the nations of early civilization be considered, but only mankind as he has attained the highest civilization at the present time.

In civilization man does not compete with plants for existence. Thorns cannot drive him from fruits, husks cannot hide nutritious seeds from his eye, shells cannot defend sweet nuts from his grasp; but he speedily destroys from the face of the earth the plants which are not of the highest value for his purpose, and he plants those that are of value and multiplies them in a marvelous manner, and by skilled culture he steadily improves their character—making the sweet sweeter, the rich richer, and the abundant more abundant.

In the higher civilization man does not compete with the beast for existence. There are no wolves or bears on our farms, there are no lions or tigers in our civilized lands, and there are no serpents in our cities. All these dwell where civilization has not yet conquered its way. Civilized man has domesticated the animal; he hives the bee for its honey, he coops the bird for its eggs, he pastures the cow for her milk, and he stables the horse that his boy may ride.

In the highest civilization the world is not crowded with human beings

---

[1] It is unfortunate for Powell that the evolutionary principle of mutualism was poorly appreciated in his day. Mutualism entails a kind of cooperation among life forms: The bee assists the flower and the flower the bee. Nature may be red in tooth and claw, but not unrelievedly so. Sometimes lending a hand can be a survival strategy, a fact that Powell might have used in support of his arguments against social Darwinism.

beyond their ability to procure sustentation; for if some hunger it is not because of the lack of the world's food, but because of the imperfect distribution of that food to all. Men are not crowded against plants, men are not crowded against beasts, and men are not crowded against one another. The land is yet broad enough for all. The valleys are not all filled, the hillsides are not all covered. The portion of the earth that is actually cultivated and utilized to supply the wants of man is very small; it compares with all the land as a garden to a plain, an orchard to a forest, a meadow to a prairie. Nature is prodigal of her gifts. The sweet air as it sweeps from zone to zone is more than enough to fan every cheek; the pure water that falls from the heavens and refreshes the earth and is again carried to the heavens on chariots of light is more than enough to refresh all mankind; the bounteous earth spread out in great continents is more than enough to furnish every man a home; and the illimitable sea has wealth for man that has not yet been touched. Thus it is that in human evolution overpopulation is not a factor, as it is in biotic evolution.[2]

In the highest civilization man does not compete with man in the struggle for existence, and thus human competition is not like biotic competition. In biotic evolution the wolf devours the fawn; but on the average he devours the weakest fawn and the strongest fawn lives to beget a fleeter race of stags; and the evolution of stag-life is accomplished by such means. But when the highwayman waylays the traveler, and there is a struggle for existence which ends in murder, no step in human evolution is accomplished thereby.

Again, in the higher civilization man does not compete with man in the

---

[2] Powell is specifically rebutting the arguments of Thomas Malthus (1766–1834), who held that population will inevitably outgrow food supply and that any hope for bettering the condition of the mass of humanity depended on the strict control of reproduction. Powell's assertion that "the bounteous earth" is more than capacious and rich enough "to furnish every man a home" contrasts interestingly with his repeated warnings about the limitations of the arid lands. Similarly, his statement that "in human evolution overpopulation is not a factor" runs counter to his own direct observations of the myriad ways in which the Paiute adapted to their environment. Powell was not always a consistent thinker, least of all when engaged in debate, as is implicitly the case here.

direct struggle for the means of existence, as does the brute. In the struggle for subsistence one ox gores another to drive him from a blade of grass; one wolf rends another to drive him from a bone. Among the animals the struggle for the means of existence is direct, rapacious, and cruel; but in civilized society man shares with his fellow-man; the poor and unfortunate are fed at the table of charity. A maimed beast is driven from the crib, but men and women will vie with one another to serve a maimed man, and one of the highest aspirations of civilized society is to dispense generous hospitality.

Vestiges of brutal competition still exist in the highest civilization; but they are called crimes, and to prevent this struggle for existence penal codes are enacted, prisons are built, and gallowses are erected. Competition in the struggle for existence is the agency by which progress is secured in plant and animal life, but competition in the struggle for existence among men is crime the most degrading. Brute struggles with brute for life, and in the aeons of time this struggle has wrought that marvelous transformation which we call the evolution of animals; but man struggles with man for existence and murder runs riot; no step in human progress is made.

That struggle for existence between man and man which we have considered and called crime is a struggle of one individual with another. But there is an organized struggle of bodies of men with bodies of men, which is not characterized as murder, but is designated as warfare. Here then we have man struggling with man on a large scale, and here it is where some of our modern writers on evolution discover the law of natural selection, the survival of the fittest in the struggle for existence, as the strongest army survives in the grand average of the wars of the world.

But when armies are organized in modern civilization the very best and strongest are selected, and the soldiers of the world are gathered from their homes in the pride of manhood and in lusty health. If there is one deformed, if there is one maimed, if there is one weaker of intellect, he is left at home to continue the stock, while the strong and courageous are selected to be destroyed. In organized warfare the processes of natural selection are reversed; the fittest to live are killed, the fittest to die are preserved; and in the grand average the weak, physically, mentally, and morally, are selected to become the propagators of the race.

In the history of civilization, justice and liberty are ofttimes established and supported, and in the history of civilization injustice and oppression are ofttimes established and supported. The right prevails, and the wrong prevails, and war continues. Whether more of good or more of evil comes to the institutions of mankind through war I know not, but this I do know, that war is not an agency for the selection of the best specimens of men to live and propagate their kind; it is not an agency for human evolution as carnivorous and predatory life is an agency for the evolution of the beast.

The struggle for existence between human individuals is murder, and the best are not selected thereby. The struggle for existence between bodies of men is warfare, and the best are not selected thereby. The law of natural selection, which Darwin and a host of others have so clearly pointed out as the means by which the progress of animals and plants has been secured, cannot be relied upon to secure the progress of mankind.[3] Whenever mankind falls under the dominion of the laws of animal evolution he himself becomes beastly and loses those attributes which make the human race immeasurably superior to the brute.

There are always too many plants born—more than can possibly live upon the face of the earth—and in the struggle for existence the majority of them are killed off; and because the best survive, all things being considered, there is progress in plant life, there is evolution. There are always too many animals born into the world—many more than can possibly live upon the face of the earth—and the vast majority are killed off; but the few best live, all things being considered, and therefore there is progress in animal-life—animal evolution. There are not too many human beings born into the world in lands of the highest civilization, because the earth is not now and never has been filled with men to the limit of its capacity; the great majority are not therefore killed off in the struggle for existence, and there is not a small remnant of the best preserved to continue human existence and secure human progress. The law of evolution which is called "survival of

---

[3] Powell here directly rebuts a central argument William Graham Sumner put forward in 1879 and repeatedly thereafter: "The law of survival of the fittest was not made by man. We can only, by interfering with it, produce the survival of the unfittest" (Bannister, *Social Darwinism*, 105.)

the fittest in the struggle for existence" does not apply to mankind. Human progress is by other agencies and in obedience to other laws.

It was a struggle through aeons of time by which lowly plants were developed into forest trees as palms and oaks and pines; but it was a struggle without consciousness and without pain. It was a struggle through aeons of time by which the lowly animal germ was transformed into the lion and eagle; and the whole course of this struggle was marked by conscious pain. In its history how many limbs have been rent asunder, how many deadly struggles have there been between beast and beast, how many stings of insects, how many serpents' fangs have dealt poison to writhing victims, how many shrieks of terror have rent the air, how many groans of pain have gone up to heaven! The progress of animal evolution has cost the world a hell of misery; and yet there are philosophers, professors in our colleges, and authors of world-renowned books who believe that the law of human evolution, the method by which man may progress in civilization, is the "survival of the fittest in the struggle for existence." . . .

From the dawn of human culture in savagery to the midday of culture in civilization, human genius has been producing many inventions for many purposes, and the good have given place to the better and the better have yielded to the best.

The sheep gathers the grass with his teeth, the ox with his tongue, and the horse with his lips, and teeth, tongue, and lips are modified and developed as these animals struggle for existence. But the savage, just a little higher than the brute, walks through natural meadows and with a stick in one hand beats the grain from the stalks of grass into a basket held in the other. Then to separate the grain from the chaff he tosses it on a tray that the passing breeze may cleanse it. Then the grain is roasted and ground between two stones, one lying on the ground, the other held in the hands— two mealing-stones; and the flour is spread on a stone and baked into a cake on the coals. So stick and basket and tray and mealing-stones and baking-stone are the implements and devices for gathering and preparing the cereal food of the savage. Then man invents a reaping-hook, then a grain-cradle, and then a reaper, and in the process of invention from the sickle to the reaper what a multitude of inventions are developed. Along this course how many tools, implements, and machines become obsolete and useless that the one great reaper may remain. Here it is that we have the survival of the fittest in the struggle for existence, and man by his genius transfers this struggle from himself to the work of his hand. The way from basket-reaping to power-reaping is long but all the steps of that way have been taken in the endeavor of mankind to secure greater happiness. . . .

Away back in early civilization they measured time by a leaking pot of water—a clepsydra, or a leaking goblet of sand—an hour-glass; and then they invented clocks with weights and pendulums, and better and better timepieces with springs and escapements, until they had modern chronometers. Along the course of the invention of these time-measuring instruments how many have fallen into disuse. Here again there has been a survival of the fittest in the struggle for existence, and this law of evolution has been transferred from man himself to the work of his hands; and the way from the hour-glass to the chronometer is long, but every step has been taken by man in the endeavor to secure greater happiness.

There is no end of illustrations to be drawn from the arts of mankind of this great fact, that man has emancipated himself from this cruel law of evolution by transferring it to the work of his hands. Man invents more devices than he can use; of the many only the few live, but these few are selected consciously and intelligently because they are the best. And all these inventions are made not because men struggle with nature for existence, but because men endeavor to secure happiness, to improve their condition; it is a conscious and intelligent effort for improvement. Human progress is by human endeavor. . . .

That men may live together in peace and render mutual assistance, there are times and conditions in society when one must command and another must obey. When in the highest civilization personal authority is no longer relegated to hereditary graded classes, men must still wield authority and men must still yield obedience, but a new principle of relegating rights and duties is established. Those to whom authority is given are selected by the people themselves, to exercise specified authority in specified cases, not by right of hereditary or conferred rank, but as constituted agents of the people. Thus the right of the elder is by long and constant endeavor transformed into the right of the noble; the right of the noble is by long and constant endeavor transformed into the right of the representative. Along the course of this evolution, from the ancient kinship clan to the modern republic, what endeavors have been made to secure peace. Ever the wise and good among mankind have been endeavoring for themselves and for their fellow-man to pass from the state of warfare of the brute to the state of social order in civilization through the invention of institutions. And how many institutions have passed away, how many have served for a time only to be replaced by those that were better. And here again among the institutions of mankind we have the survival of the fittest in the struggle for existence by human selection.

In the developing complexities of human industry and in the growing interdependence of men in society, those rights and duties that arise from the creation and accumulation of property constantly multiply, and the adjustment of these rights and duties is constantly in progress. It is thus that institutions for the regulation of conduct in relation to property are ever multiplying; in this manner new rights are established and new duties imposed. . . .

All of this statement may be summarized in this manner: Man does not compete with plants and animals for existence, for he emancipates himself from that struggle by the invention of arts; and again, man does not compete with his fellow-man for existence, for he emancipates himself from the brutal struggle by the invention of institutions. Animal evolution arises out of the struggle for existence; human evolution arises out of the endeavor to secure happiness: it is a conscious effort for improvement in condition.

But the arts and institutions alone have not secured the evolution of mankind, for they have been powerfully aided by two other classes of human inventions, namely, linguistics and opinions; and the part which they have taken must be mentioned. The same struggle for existence and the same survival of the fittest by human selection which have been found among arts and again among institutions may be discovered among languages and linguistic methods and devices. By human endeavor man has created speech by which he may express his thought. The nightingale sings to his mate; the poet sings to mankind. The song of the nightingale dies with the passing of the zephyr; the song of the poet lives for ages. And in all linguistic inventions through the coining of words and devising of grammatic methods and the invention of alphabets and of printing and of telegraphs and of telephones, invention has struggled with invention for existence, and the many have been sacrificed that the few—the best—might remain. Through all the time of human history man has sought to communicate his thoughts to man; and wherever man has lived he has endeavored to communicate with his fellow. In every valley there has been a village or a town or a city where men have endeavored to communicate their thoughts; and everywhere the orator has endeavored to plead with greater eloquence, everywhere the teacher has endeavored to expound with greater clearness, everywhere the advocate has endeavored to persuade with greater force, everywhere the philosopher has endeavored to set forth the greater truth, everywhere the poet has endeavored to sing with greater beauty, and everywhere the lover has endeavored to tell a more tender story; and out of this endeavor in all lands and in all time the unorganized language of savagery has been developed into the languages of modern civilization; and all this progress, all this evolution, is by human endeavor, and in it natural selection, as that term is understood in biology, has played no part.

Along the course of human progress opinions have been changing. The cruelty of nature in biotic evolution has been set forth. In this figure of speech Nature is personified, and if we still personify nature, to the savage man Nature is ever a deceiver and a cheat.

Nature tells the savage that the earth is flat, over which the sky is arched as a solid dome; then Nature tells the savage that the sun travels over the flat earth and under the sky of ice by day from east to west; then Nature tells the savage that the rain comes from the melting of the ice of the sky. Many, strange, foolish, and false are the stories that Nature tells to the untutored savage. Nature is the Gulliver of Gullivers, the Munchausen of Munchausens.[4] Nature teaches men to believe in wizards and ghosts; Nature fills the human mind with foolish superstitions and horrible beliefs. The opinions of the natural man fill him with many fears, give him many pains, and cause him to commit many crimes. Out of all these savage superstitions man has travelled a long way into the light of science. And how shall the opinions of modern civilization be characterized. And who can tell how the knowledge of the highest civilization transcends the knowledge of the lowest savagery.

The problems of the universe are not all solved, though savage philosophers and barbaric philosophers and civilized philosophers, in all lands and in all times, have sought for the truth. And so opinions have been changing, old opinions have died and new opinions have been born, and philosophies have struggled for existence as man has endeavored to learn; and with man forever the struggle to know has been the endeavor to secure happiness, for truth is good and wisdom is joy.

I have thus set forth the four great classes of activities by means of which human progress has been accomplished. Human evolution is the result of the development of the arts, institutions, linguistics, and opinions; and these are the product of man's genius; they are human inventions—the result of his endeavor to secure happiness; and in all there is no "natural selection," but only human selection. . . .

All honest men are working for other men. If a man works exclusively for himself he is a counterfeiter or a forger or a sneak-thief, or perchance a highwayman. All love of industry, all love of integrity, all love of kindred, all love of neighbor, all love of country, and all love of humanity is expressed in labor for others. For the service thus performed a right to a reward is acquired, and he for whom the service is performed has imposed upon him

---

[4] "Baron Munchausen's Narrative of His Marvelous Travels and Campaigns in Russia" first appeared in London in 1785. This collection of tall tales was the unattributed work of Rudolph Erich Raspe; with expansions by G. A. Burger in 1786 and 1788, it grew into an operatic celebration of fantastical lying. Lemuel Gulliver, meanwhile, is the central character and falsely purported author of "Travels into Several Remote Nations of the World" (1726), which is better known as the classic satire *Gulliver's Travels* by Jonathan Swift.

the duty to render the reward, and the service is rendered in the hope of the reward. Everywhere in civilized society men are thus working for others. Every man in all the years of his labor toils for his fellow-man and the practice is universal among all honest civilized men, and lasts from generation to generation; and universal practice is gradually becoming crystallized into universal habit. One man is trying to construct better houses for his neighbors, another man is trying to make better shoes for his neighbors, another man is trying to make better laws for his neighbors, and another man is trying to make better books for his neighbors. Every man is thus forever dwelling upon the welfare of his neighbors and making his best endeavor for his good, and thus the habit grows from generation to generation, until at last some men even forget that there is reward for service, and labor for their fellow-men because they love their fellow-men. They witness the disabilities of the maimed and they build them homes, for the sick they build hospitals, for the insane they erect asylums, for the blind and deaf they establish schools, for the ignorant they furnish instruction, for the wronged they enact laws and demand justice, for the enslaved they fight for liberty; and thus it is that while a large share of the energy of mankind is exerted to benefit his neighbor in hope of reward, another and larger share is exerted for the love of his child, for the love of his brother, for the love of his country, for the love of mankind. Man toils for the reward which must be tendered by his neighbor; but, more than all and higher than all, man toils for others without hope of reward from them, but receiving the bounteous reward of an approving conscience. . . .

As man toils for his neighbor in the hope of material reward, there arises a condition of activities which in political economy is called "competition," but which differs altogether from that competition which obtains among plants and animals. Let us see what this so-called competition is.

Economic competition exists among men engaged in the same vocation and who may desire to render service to the same persons. It is rivalry to render a service to others that the reward for that service may be received.

Economic competition has or may have two factors: emulation and antagonism. By emulation is meant the strife between men for greater excellence—to perform better service for their fellow-men. By antagonism is meant strife in which man endeavors to injure his rival that he may himself succeed. We have, then, emulative competition and antagonistic competition. Emulative competition results in human progress; antagonistic competition results in human retrogression.

Let us see in what manner this economic competition works out its results among certain great classes that exist in the highest industrial society.

Artists compete with one another. Every painter strives to make the best landscape, the best historic group, the best general composition, or the best portrait. This is emulation. But in companionship with it, other endeavors may arise. The painters may organize leagues or schools to instruct one another; and there ofttimes grows up among these artists, who create with the brush, such an appreciation of common interest in art as leads to great mutual help, and a comradeship that inspires to the best endeavor. Such generous emulation and all its products are in the life of human progress.

But jealousies grow, and of these are born unjust criticism, carping detraction, and vile slander; and such antagonistic competition is akin to that existing among the brutes and leads to no progress among mankind. So generous emulation among sculptors, generous emulation among musicians, generous emulation among poets, and generous emulation among romancers leads to human progress.

In art emulative competition is advantageous to the artists themselves in enabling them to derive aid and inspiration from one another; and every success in art creates among laymen an appreciation and love of art in every way beneficial to the artist himself.

There is not in civilized society a limited demand for works of art of such a nature that what one man produces is a prohibition upon the production of some other man. In any given community there is not a natural demand or desire for 999 works of art. The natural man—the man in his ignorance—spurns them all. It is the cultured man only that loves art, and the culture which brings appreciation and love of art arises from the ethical training which works of art themselves give. In art, demand does not create supply, but supply creates demand. It is thus that every broad-minded artist rejoices at the success of his brother.

Again, among the professional classes both emulative and antagonistic competition develops. The emulative competition is a powerful factor in human progress, but the antagonistic competition ever works to thwart and retard it, and to avoid the evils of antagonistic competition and transform it into emulative competition, associations or societies are organized. Thus the physicians of a city organize an association and meet to discuss questions relating to the principles and application of their science, to promote personal intercourse and friendship, and to regulate rates of compensation.

A large part of the people of civilization are engaged in agriculture, and agricultural products are poured into the common reservoir—the markets of the world. The clientage of the farmer is thus large and indefinite. He is not striving to serve his neighbor, Jones, but to serve the world. The farmers, too, are of a great number—that is, there are many servants. For these reasons a farmer does not compete with his neighbor or with any number of specified or known persons, but his competition is with the whole body of farmers. For this reason the spirit of antagonistic competition is never born; the competition of the farmer with the farmer is purely emulative. So farmers organize agricultural societies, and establish agricultural colleges, and support agricultural papers, and by every possible agency diffuse knowledge among themselves relating to their vocation; and emulative competition is wholly an agency for progress. It is for this reason that the asperities of industrial life scarcely exist among agricultural people.

More than any other man in modern society, the farmer works for himself, as a part of the products of the farm are consumed at the farmer's table. But it is probable that what is thus gained in emancipation from competition is offset in the loss of interdependence arising therefrom.

Another large body of people are engaged in mining, manufacturing, and transporting industries, and these classes may be thrown together for the purposes of this exposition. In a broad way they may be separated into employers and employed. The employers compete with one another in the sale of their productions, and it is both emulative and antagonistic; but that the evils of antagonistic competition may be avoided, each class of employers is gradually organizing corporations or trusts. It will thus be seen that antagonistic competition is obviated by organization and consolidation; but by organization and consolidation cumulative competition is also avoided, for the managers of business enterprises no longer compete for business, but distribute business by convention. And in the same manner they repeal the law of competition in the labor market; they seek by convention to establish rates of wages.

The employees in these productive industries of mining, manufacturing and transporting naturally compete with one another in two ways: To secure employment they endeavor to render their labor more efficient by skill and industry, and they offer to labor for smaller wages. The method of competition by the improvement of labor is emulative; the method of competition by cheapening labor is antagonistic. In all civilized society there is no competition so direful in its results, so degrading to mankind, as that which is produced among the employees of these classes who compete for employment by cheapening labor, for it results in overwork, which is brutalizing, and in want, which is brutalizing, and the abolition of this form of competition is one of the great questions of the day. In order to avoid the evil to themselves of antagonistic competition they organize themselves into societies, which may in general be termed labor unions. In this class, composed of the people who engage in the productive industries of mining, manufacturing, and transporting, emulative competition and antagonistic competition are both of great potency, and employers and employed alike resort to organization for self-protection, and the conflict is irrepressible. Considering the two classes of employers and employees, the growth of organization is steadily resulting in the destruction of antagonistic competition, but to a large extent it also results in the destruction of emulative competition. The great problem in industrial society today is to preserve emulative competition and to destroy antagonistic competition. Few comprehend the philosophy of the industrial struggle of today. The professional classes have already solved the problem for themselves, and they stand aloof and deplore the struggle, but they should learn this lesson from history, that when wrongs arise in any class of society those wrongs must ultimately be righted, and so long as they remain, the conflict must remain, and when the solution comes not by methods of peace it comes by war.[5]

Injustice is a strange monster. Let any body of people come to see that injustice is done them in some particular, though it may be one which affects their welfare but to a limited degree, they will dwell upon it, and discuss it, and paint its hideous form from one to another, until the spectre of that injustice covers the heavens, and gradually to it the people will attribute all their evils. If a body of laborers receive unjust reward for their toil they will dwell upon the evil so long, so often, and kindle their passions to such a height that they will at last attribute to the failure of receiving a modicum of reward for their toil, all the evils of their own improvidence; all the evils of their own intemperance; all the evils of their own lust; and if fire and flood come, the very evils of unavoidable misfortune will be attributed to the injustice of unrequited toil. Injustice is of such a nature that it must be destroyed by society or it will destroy society. We dare not contemplate its existence with equanimity, for "Behold how great a matter a little fire kindleth."

Among the exchanging classes, economic competition exists in the plenitude of its power, and it is both emulative and antagonistic. The friction of exchange is great, but all emulative competition in this department of human industry lessens this friction and results in a benefit to the consumers of the world. It is in the antagonistic competition that evil arises, and here its proportions are enormous and its forms Protean.[6]

---

[5] Widespread unemployment and destitution accompanied the economic depression of the 1870s, and intense labor conflicts became increasingly common in the industrializing United States. In the 1880s violent strikes on midwestern railroads culminated in a bloody clash in Haymarket Square in Chicago in May 1886 that left seven police and four civilians dead and scores wounded. In that year, the American Federation of Labor, a national union representing many trades, took form with Samuel Gompers as president. The "war" to which Powell refers continued to intensify in the years after publication of this article. In 1892 the Homestead Steel strike resulted in the call-up of the Pennsylvania National Guard. In the following year, the Pullman strike effectively brought rail traffic to a halt throughout much of the nation.

[6] Proteus was a Greek god who could assume multiple forms—a lion, water, a tree, etc.

No better illustration of this subject can be found than the method of antagonistic competition that exists in advertising. The honest system of advertising should be but a simple announcement of the offer of goods for the information of those who desire to purchase, in such a manner that they may by seeking find. But in advertising as it now exists, exaggeration is piled on exaggeration, and falsehood is added to falsehood. The world is filled with monstrous lies, and they are thrust upon attention by every possible means. When a man opens his mail in the morning the letter of his friend is buried among these advertising monstrosities. They are thrust under street-doors, and they are offered as you walk the streets. When you read the morning and evening papers, they are spread before you with typographic display; they are placed among the items you desire to read, and they are given false headings, and they begin with decoy paragraphs. They are pasted on walls and on the fences and on the sidewalks and on bulletin boards, and the barns and house-tops and fences of all the land are covered with them, and they are nailed to the trees and painted on rocks. Thus it is that the whole civilized world is placarded with lies, and the moral atmosphere of the world reeks with the foul breath of this monster of antagonistic competition.

Illustrations of the two kinds of economic competition may be found on every hand, for the world is full of them, and one must be nurtured and the other must be destroyed.

Many years ago in the land of our forefathers, that beautiful isle of Great Britain, there were large tracts of unappropriated land that from time immemorial had been left as commons, where the cottager might pasture his cow and where his children might play at will; but the time came when all this land was of great value, and the philanthropists said: "Let us distribute these tracts among the agricultural laborers and give to every one a home"; for there was land enough to secure this purpose. But there are always great philosophers in this world, and one of these great philosophers[7] discovered a fundamental principle, a major premise; it was this: that you cannot improve the condition of the poor by giving them homes, for if you do they

---

[7] Thomas Malthus; see note 2.

will multiply so fast in the wantonness of lust that their last condition is worse than their first. And so as the nobles wanted the land, the great estates were made greater by enclosing the commons. At last humanity has learned that wanton extravagance of life is cured by elevating the poor to a higher condition, where they speedily learn the principles of prudential reproduction, and to-day in that land statesmen and scholars are devising the means by which those great estates may yet be distributed among the poor. In that elder day the problem was of simple solution, but to-day it is the greatest institutional enterprise that civilized man has ever undertaken.

Then still later in that beautiful isle the employing classes were more powerful than the employed, and they reduced the wages of the laborer to the lowest pittance and pleaded in justification the sanction of the immutable law of competition. Then philanthropists sought to devise methods by which the laborers might receive better wages, and they raised great funds by taxes, and taught the laborers to eke out a miserable existence on the fruits of parish-doled charity. Then another great philosopher[8] arose among the people and propounded a new fundamental principle, a major premise; it was this: At any time in the history of a nation there is only a definite amount of money which can be used in the employment of labor, and if you give more to one you must give less to another; it is therefore impossible in the aggregate to increase wages unless, indeed, you resort to the method of killing off a part of the laborers. And so wages were reduced still lower. Strikes and riots were organized, and the beautiful isle was threatened with anarchy, till at last employers were compelled to yield a greater share of justice. And now another philosopher[9] has arisen in the world, and he has discovered another fundamental principle, a major premise; that human progress is by the survival of the fittest in the struggle for

---

[8]David Ricardo (1772–1823) propounded the Iron Law of Wages: "The natural price of labor is that price necessary to enable the laborers one with another to subsist and to perpetuate their race without either increase or diminution."

[9]Herbert Spencer (1820–1903), who coined the term *survival of the fittest,* advocated the preeminence of the individual over larger social units and an extreme form of social and economic laissez-faire that became known, usually pejoratively, as social Darwinism.

existence; that the fittest may survive, the unfit must die. Then let the poor fall into deeper degradation, then let the hungry starve, then let the unfortunate perish, then let the ignorant remain in his ignorance—he who does not seek for knowledge, himself is not worthy to possess knowledge, and the very children of the ignorant should remain untaught, that the sins of the fathers may be visited upon the children. Let your government cease to regulate industries, and instead of carrying the mails let them erect prisons; let government discharge their state-employed teachers and enlist more policemen. And they establish journals to advocate these principles, and edit papers to advocate these principles, and they have become the most active propagandists of the day, and the millions are shouting, "Great is philosophy and great are the prophets of philosophy."

Thus it is that fundamental principles, "major premises," are discovered to justify injustice, and yet forever man is endeavoring to establish justice. How this shall be done I know not, but I have faith in my fellow-man, towering faith in human endeavor, boundless faith in the genius for invention among mankind, and illimitable faith in the love of justice that forever wells up in the human heart.

The following list is necessarily incomplete, for Powell was an astonishingly prolific writer and his contributions, especially to government documents, are scattered among many volumes. Prominent among these are the annual reports of the United States Geological Survey and the Bureau of Ethnology during the years Powell headed them (1881–1894 and 1879–1902, respectively). The prospective researcher will note, however, that for a variety of reasons the publication of annual reports was often delayed well past the period to be covered. For example, the *Seventh Annual Report of the Bureau of Ethnology,* which reported on work performed in 1885 and 1886 (and contained Powell's important treatise on Indian languages, "Indian Linguistic Families of Indians North of Mexico"), did not appear until 1891.

Two more government reports by Powell are particularly and deservedly famous: *Exploration of the Colorado River of the West and Its Tributaries* (1875) and *Report on the Lands of the Arid Region* (1878 and 1879). At least one other document is no less significant as a statement of his ideas and recommendations for the settlement of the arid lands of the West and the administration of their resources. This is the second annual report of the Irrigation Survey, which is included as "Part II: Irrigation" in the *Eleventh Annual Report of the United States Geological Survey,* 1890–91 (Washington, D.C.: Government Printing Office, 1891). The report includes most of the testimony Powell presented to the House Select Committee on Irrigation of the Arid Lands during a marathon series of seven hearings through the spring of 1890 (February 27; March 1, 13, 15, 27; and April 17 and 24). In prepared statements and in exchanges with committee members Powell ably presents his vision for the West and displays his encyclopedic knowledge of individual watersheds throughout the region. Transcripts of these important hearings are also included, with additional testimony from Powell, Clarence Dutton, Anson Mills, and others, in Appendix B ("Irrigation in the United States") of House Report 3767, "Ceding the Arid Lands to the States and Territories," 51st Cong. 2nd Sess., pp. 1–202 (1891) Serial No. 2888. Analogous Senate documents may be found in Senate Report 928, fifty-first Congress, First Session (1890) Serial No. 2708. This is the report of the Special Committee of the U.S. Senate on the Irrigation and Reclamation of Arid Lands, which was chaired by Senator William ("Big Bill") Stewart of Nevada. Powell's principal statement appears in Part IV (pp. 5–95 and 151–204). Frequent interruptions by Senator Stewart, who by 1890 had become Powell's most vehement congressional adversary, deny Powell's Senate testimony the lucidity of his exposition before the House committee. His command of his material and of himself, however, remains extraordinary. Readers or researchers seeking the fullest expression of Powell's views on western lands should not neglect these records of Powell's virtuoso performances in the winter and spring of 1890.

The following list is presented chronologically to reflect the growth and pattern of Powell's interests and the arc of his productivity.

1867–1875

Powell, John Wesley. 1867. *Scientific Expedition to the Rocky Mountains. Preliminary Report of Professor J. W. Powell.* Proclamation 111. Illinois State Board of Education, 9–13.

———. 1947–49. *Records of the Powell Colorado River Expeditions* [includes letters to the *Chicago Tribune* via wilderness post, journals of both Colorado River expeditions, and much else]. *Utah Historical Quarterly* 15–17.

———. 1870. "Major J. W. Powell's Report on His Explorations of the Rio Colorado in 1869." Included as Appendix D in William A. Bell, *New Tracks in North America.* New York: Scribner, Welford & Co. (Reissued, Albuquerque: Horn and Wallace, 1965.)

———. 1872. *Report of the Survey of the Colorado River of the West.* March 25, 1872. Forty-Second Congress, Second Session. H.R. Misc. Document 173.

———. 1873. *Report of the Survey of the Colorado River of the West.* January

17, 1873. Forty-Second Congress, Third Session. H.R. Misc. Document 76.

———. 1873. "Some Remarks on the Geological Structure of a District of Country Lying to the North of the Grand Cañon of the Colorado." *American Journal of Science and Arts,* Third series, 5: 456–65.

———. 1874. *Geographical and Geological Surveys West of the Mississippi.* Forty-Third Congress, First Session. H.R. Report 612.

———. and G. W. Ingalls. 1874. *Report of Special Commissioners on the Condition of the Ute Indians of Utah; the Paiutes of Utah, Northern Arizona, southern Nevada, and southeastern California; the Go-si-utes of Utah and Nevada; the Northwestern Shoshones of Idaho and Utah; and the Western Shoshones of Nevada; and Report Concerning Claims of Settlers in the Mo-a-pa Valley, Southeastern Nevada.* Washington, D.C.: Government Printing Office.

———. 1874. *Report on the Survey of the Colorado River of the West.* April 30, 1874. Forty-Third Congress, First Session. H.R. Misc. Document 265.

———. 1874. "Remarks on the Structural Geology of the Colorado of the West." *Bulletin of the Philosophical Society* (Washington, D.C.) 1: 48–51.

———. 1875. "An Overland Trip to the Grand Cañon." *Scribner's Monthly* 10: 659–78.

———. 1875. *Exploration of the Colorado River of the West and Its Tributaries.* Forty-Third Congress, First Session. H.R. Misc. Document 300.

———. 1875. "The Cañons of the Colorado." *Scribner's Monthly* 9: 293–310, 394–409, 523–37.

———. 1875. "Physical Features of the Colorado Valley." *Popular Science Monthly* 7: 385–99, 531–42, 670–80.

1876–1880

———. 1876. "The Ancient Province of Tusayan." *Scribner's Monthly* 11: 193–213.

———. 1876. *Report on the Geology of the Eastern Portion of the Uinta Mountains and a Region of the Country Thereto.* Washington, D.C.: U.S.G.S. Territories.

———. 1877. "Types of Orographic Structure." *American Journal of Science and Arts,* Third series, 12: 414–28.

———. 1877. *Introduction to the Study of Indian Languages.* Washington, D.C.: Government Printing Office, 1877. Second edition (revised), Washington, D.C.: Government Printing Office, 1880.

———. 1878. "A Discourse on the Philosophy of the North American Indians." *Journal of the American Geographical Society of New York* 8: 251–68.

———. 1878. *Report on the Lands of the Arid Region of the United States, with a More Detailed Account of the Lands of Utah.* Forty-Fifth Congress, Second Session. H.R. Exec. Document 73.

———. 1878. *Report on the Methods of Surveying the Public Domain to the Secretary of the Interior at the Request of the National Academy of Sciences.* Washington, D.C.: Government Printing Office.

———. 1879–1880. "Mythologic Philosophy." *Popular Science Monthly* 15: 795–808; 16: 56–66.

———. 1880. "Anthropology" (a contribution to "General Notes"). *The American Naturalist* 14: 603–5.

———. 1880. "Sketch of Lewis H. Morgan, President of the American Association for the Advancement of Science." *Popular Science Monthly* 18: 114–21.

1881–1885

———. 1881. "On the Evolution of Language, as Exhibited in the Specialization of the Grammatic Processes, the Differentiation of the Parts of Speech and the Integration of the Sentence; from a Study of Indian languages." In *First Annual Report of the Bureau of American Ethnology,* 1–16. Washington, D.C.: Government Printing Office.

————. 1881. "Sketch of the Mythology of the North American Indians." In *First Annual Report of the Bureau of American Ethnology*, 17–56. Washington, D.C.: Government Printing Office.

————. 1882. "Outlines of Sociology." *Transactons of the Anthropological Society of Washington* 1: 106–29.

————. 1882. "Darwin's Contributions to Philosophy." *Proceedings of the Biological Society of Washington* 1: 60–70.

————. 1883. "Human Evolution." *Transactions of the Anthropological Society of Washington* 2: 176–208.

————. 1883. "Review of Ward's 'Dynamic Sociology.'" *Science* 2: 45–49, 105–8, 171–74, 223–26.

————. 1884. "The Three Methods of Evolution." *Bulletin of the Philosophical Society of Washington* 6: 27–52.

————. 1884. "On the State of the Interior of the Earth." *Science* 3: 480–82.

————. 1884. "On the Fundamental Theory of Dynamic Geology." *Science* 3: 511–13.

————. 1884. *Address Delivered at the Inauguration of the Corcoran School of Science and Arts, in the Columbia University, Washington, D.C., October 1, 1884* (pamphlet).

————. 1885. "From Savagery to Barbarism." *Transactions of the Anthropological Society of Washington* 3: 173–96.

————. 1885. "The Organization and Plan of the United States Geological Survey." *American Journal of Science*, Third edition, 29: 93–102.

————. 1885. "The Administration of the Scientific Work of the General Government." *Science* 5: 51–55.

————. 1885. "The Larger Import of Scientific Education." *Popular Science Monthly* 26: 452–56.

1886–1890

————. 1886. *Testimony Before a Joint Commission to Consider the Present Organization of the Signal Service, Geological Survey, Coast and Geodetic Survey, and the Hydrographic Office of the Navy Depart-*

ment, with a View to Secure Greater Efficiency and Economy of Operation. Forty-Ninth Congress, First Session. Senate Misc. Document 82.

———. 1887. "The Cause of Earthquakes." *Forum* 2: 370–91.

———. 1888. "From Barbarism to Civilization." *American Anthropology* 1: 97–123.

———. 1888. "Competition as a Factor in Human Evolution." *American Anthropology* 1: 297–323.

———. 1888. "Trees on Arid Lands." *Science* 12/297: 170–71.

———. 1888. "The Laws of Hydraulic Degradation." *Science* 12: 229–33.

———. 1888. *Methods of Geologic Cartography in Use by the United States Geological Survey.* Berlin: International Geological Congress, Third Session, 221–40.

———. 1888. "The Personal Characteristics of Professor Baird." *Bulletin of the Philosophical Society of Washington* 10: 71–77.

———. 1889. "Address to the North Dakota Constitutional Convention, August 5, 1889." In *Reclamation Era* 26/9: 201–2, 1936.

———. 1889. "Address to the Montana Constitutional Convention, August 9, 1889." In *Proceedings and Debates of the Constitutional Convention.* Helena, Montana: State Publishing Company, 1921, 820–23.

———. 1889. "The Lesson of Conemaugh." *North American Review* 149: 150–56.

———. 1890. "The Irrigable Lands of the Arid Region." *Century Magazine* 39: 766–76.

———. 1890. "The Non-Irrigable Lands of the Arid Regions." *Century Magazine* 39: 915–22.

———. 1890. "Institutions for the Arid Lands." *Century Magazine* 40: 111–16.

———. 1890. *Irrigation and Reclamation of Public Lands.* Fifty-First Congress, First Session. Senate Report 1466.

———. 1890. "Evolution of Music from Dance to Symphony." *Proceedings of the American Association for the Advancement of Science,* Thirty-Eighth Meeting, 1–21, Washington, D.C.

1891–1895

———. 1891. "The New Lake in the Desert." *Scribner's Magazine* 10: 463–68.

———. 1891. *Indian Linguistic Families of America North of Mexico.* Bureau of American Ethnology, Seventh Annual Report, 1–142, Washington, D.C.

———. 1891–1892. "National Agencies for Scientific Research." *Chautauquan* 14: 37–42, 160–65, 291–97, 422–25, 545–49, 668–73.

———. 1892. "Our Recent Floods." *North American Review* 155: 149–59.

———. 1893. "History of Irrigation." *The Independent* 45/2318 (May 4, 1893): 593–95.

———. 1893. "The Mineral Exhibits at Chicago" [unsigned]. *British Trade Journal* 31: 520–22.

———. 1893. "Are Our Indians Becoming Extinct?" *Forum* 15: 343–54.

———. 1893. Address and comments. *International Irrigation Congress Official Proceedings* 2: 107–16.

———. 1893. "Simplified Spelling." *American Anthropology* 6: 193–95.

———. 1894. "The Water Supplies of the Land and Region." *Irrigation Age* 6: 54–65.

———. 1894. "Ownership of Land in the Region." *Irrigation Age* 6: 143–49.

———. 1894. "The North American Indians." In Nathaniel S. Shaler, *The United States of America: A Study of the American Commonwealth.* Vol. 1: 190–272. New York, NY.

———. 1894. "On the Nature of Motion." *The Monist* 5: 55–64.

———. 1894. "Immortality" [verse]. *Open Court* 8: 4335–37.

———. 1895. *Canyons of the Colorado.* Meadville, Pa.: Flood and Vincent.

———. 1895. "Physiographic Processes." *National Geographic Monthly* 1: 1–32.

———. 1895. "Physiographic Features." *National Geographic Monthly* 1: 33–64.

———. 1895. "Physiographic Regions of the United States." *National Geo-*

graphic Monthly 1: 65–100.

———. 1895. "Proper Training and the Future of the Indians." *Forum* 18: 622–29.

———. 1895. "The Soul" [verse]. *The Monist* 5/3: 1–16.

1896–1901

———. 1896. "Relation of Primitive Peoples to Environment, Illustrated by American Examples." *Annual Report of the Board of Regents of the Smithsonian Institution,* July 1895, 625–37, Washington, D.C.

———. 1896. "On Primitive Institutions." *Report of the 19th Annual Meeting of the American Bar Association,* 573–93.

———. 1896. "Seven Venerable Ghosts." *American Anthropology* 9: 67–91.

———. 1896. "James Dwight Dana." *Science* (New Series), 3: 181–85.

———. 1897. "On Regimentation." *Fifteenth Annual Report of the Bureau of American Ethnology,* civ–cxxi. Washington, D.C.

———. 1898. "An Hypothesis to Account for the Movement in the Crust of the Earth." *Journal of Geology* 6: 1–9.

———. 1898. "The Five Categories of Human Activities: Esthetology, Technology, Sociology, Philology, and Sophiology." *Seventeenth Annual Report of the Bureau of American Ethnology,* xxxvii–xxxviii. Washington, D.C.

———. 1898. "The Evolution of Religion." *The Monist* 8: 183–204.

———. 1898. *Truth and Error or the Science of Intellection.* Chicago: Open Court.

———. 1899. "Esthetology, or the Science of Activities Designed to Give Pleasure." *American Anthropology* (New Series), 1: 1–40.

———. 1899. "Technology, or the Science of Industries." *American Anthropology* (New Series), 1: 319–49.

———. 1899. "Sociology, or the Science of Institutions." *American Anthropology* (New Series), 1: 475–509, 695–744.

———. 1900. "The Lessons of Folklore." *American Anthropology* (New Series), 2: 1–36.

———. 1900. "Philology, or the Science of Activities Designed for Expression." *American Anthropology* (New Series), 2: 603–37.

———. 1901. "Sophiology, or the Science Designed to Give Instruction." *American Anthropology* (New Series), 3: 51–79.

◇

Alexander, J. A. *The Life of George Chaffey.* New York: Macmillan, 1928.

Allen, Craig D., ed. *Fire Effects in Southwestern Forests: Proceedings of the Second La Mesa Fire Symposium.* General Technical Report, RM-GTR-286. Fort Collins, Colo.: USDA Forest Service, 1996.

Bannister, Robert C. *Social Darwinism: Science and Myth in Anglo-American Social Thought.* Philadelphia: Temple University Press, 1979.

Bartlett, Richard. *Great Surveys of the American West.* Norman: University of Oklahoma Press, 1962.

Bogan, Michael A., Craig D. Allen, Esteban H. Muldavin, Steven P. Platania, James N. Stuart, Greg H. Farley, Patricia Melhop, and Jayne Belnap. "Southwest." In M. J. Mac, P. A. Opler, C. E. Puckett Haecker, and P. D. Doran, eds. *Status and Trends of the Nation's Biological Resources.* Reston, Va.: United States Geological Survey, 1998.

Brew, J. O. "Hopi Prehistory and History to 1850." In Alfonso Ortiz, volume ed., *Handbook of North American Indians,* vol. 9: "Southwest." Washington, D.C.: Smithsonian Institution, 1979.

Connelly, John C. "Hopi Social Organization." In Alfonso Ortiz, volume ed. *Handbook of North American Indians,* vol. 9: "Southwest." Washington, D.C.: Smithsonian Institution, 1979.

Cooley, John, comp. and ed. *The Great Unknown: The Journals of the Historic First Expedition down the Colorado River.* Flagstaff, Ariz.: Northland, 1988.

Darrah, William Culp. *Powell of the Colorado.* Princeton, N.J.: Princeton University Press, 1951.

Devlin, Bernie, Stephen E. Fienburg, Daniel P. Resnick, and Kathryn Roeder, eds. *Intelligence, Genes, and Success: Scientists Respond to the Bell Curve.* New York, NY: Copernicus Books, 1997.

Diamond, Jared. *Guns, Germs, and Steel*. New York: W.W. Norton, 1997.

Elliott, Melinda. *Great Excavations: Tales of Early Southwestern Archaeology, 1888–1939*. Santa Fe, N.M.: School of American Research, 1995.

Fischer, Claude S., et al., eds., *Inequality by Design: Cracking the Bell Curve Myth*. Princeton, N.J.: Princeton University Press, 1996.

Gannett, Henry. "Do Forests Influence Rainfall?" *Science* 11 (January 6, 1888): 3–5.

———. "The Influence of Forests on the Quantity and Frequency of Rainfall." *Science* 12 (November 23, 1888): 242–44.

Gilbert, Grove Karl. "John Wesley Powell" [obituary], *Science* 16/406 (October 10, 1902): 567.

Goddard, Ives. "The Classification of the Native Languages of North America." In Ives Goddard, volume ed. *Handbook of the Indians of North America*, vol. 17: "Languages." Washington, D.C.: Smithsonian Institution, 1996.

Goetzmann, William. *Exploration and Empire*. New York: Norton, 1966.

Herrnstein, Richard J. and Charles Murray. *The Bell Curve: Intelligence and Class Structure in American Life*. New York: Free Press, 1994.

Hibbard, Benjamin H. *A History of the Public Land Policies*. Madison,1 1924, 1965.

Hill, Mary. *California Landscape: Origin and Evolution*. Berkeley: University of California Press, 1984.

Hofstadter, Richard. *Social Darwinism in American Thought*. Introduction by Eric Foner. Boston: Beacon Press, 1992. (Originally published 1944; Beacon Press first edition 1955.)

International Irrigation Congress. *Official Proceedings* 2 (1893): 107–16.

Kelly, Isabel T. and Catherine S. Fowler. "Southern Paiute." In Warren D'Azvedo, volume ed. *Handbook of North American Indians*, vol. 11: "Great Basin." Washington, D.C.: Smithsonian Institution, 1986.

Kemmis, Daniel. *Community and the Politics of Place*. Norman: University of Oklahoma Press, 1990.

Kincheloe, Joe L., Shirley R. Steinberg, and Aaron D. Gresson, eds. *Measured Lies: The Bell Curve Examined.* New York: St. Martin's Press, 1996.

King, Clarence. "On the Discovery of Actual Glaciers on the Mountains of the Pacific Slope." *American Journal of Science and Arts* 1/3 (March 1871): 157–67.

Larsen, Wesley P. "The 'Letter' or Were the Powell Men Really Killed by Indians." *Canyon Legacy* 17 (Spring 1993): 12–19.

McGinnis, Michael Vincent, ed. *Bioregionalism.* London: Routledge, 1998.

Nabokov, Peter. *Indian Running: Native American History and Tradition.* Santa Fe, N.M.: Ancient City Press, 1987.

Pinchot, Gifford. *Breaking New Ground.* Commemorative edition. Introduction by Char Miller and V. Alaric Sample. Washington, D.C.: Island Press, 1998. (Originally published by Harcourt, Brace, 1947.)

Pisani, Donald J. "Forests and Reclamation, 1891–1911." *Forest and Conservation History* 37/2 (April 1993): 68–79.

———. *To Reclaim a Divided West: Water, Law, and Public Policy, 1848–1902.* Albuquerque: University of New Mexico Press, 1992.

Pyne, Stephen J. *Fire in America.* Princeton, N.J.: Princeton University Press, 1982.

Rabbitt, Mary C. "John Wesley Powell: Pioneer Statesman of Federal Science." In *The Colorado River Region and John Wesley Powell.* Geological Survey Professional Paper 669. Washington, D.C.: U.S. Government Printing Office, 1969.

Reisner, Marc. *Cadillac Desert: The American West and Its Disappearing Water.* New York: Penguin, 1986.

Rothman, Hal, ed. *Reopening the American West.* Tucson: University of Arizona Press, 1998.

Sale, Kirkpatrick. *Dwellers in the Land: The Bioregional Vision.* Athens: University of Georgia Press, 2000. (Originally published by Sierra Club Books, 1998.)

Service, Elman R. "Cultural Evolution." In David L. Sills, ed. *International Encyclopedia of the Social Sciences,* vol. 5. New York: Macmillan and the Free Press, 1968, 221–28.

Sexton, W. T., A. J. Malk, R. C. Szaro, and N. C. Johnson. *Ecological Stewardship: A Common Reference for Ecosystem Management.* 3 vols. Oxford, U.K.: Elsevier, 1999.

Shaul, D. Leedom. "Linguistic Natural History: John Wesley Powell and the Classification of American Languages." *Journal of the Southwest* 41/3 (Autumn 1999): 297–310.

Smith, Henry Nash. "Rain Follows the Plow: The Notion of Increased Rainfall for the Plains." *Huntington Library Quarterly* 10 (February 1947): 169–93.

———. *Virgin Land: The American West as Symbol and Myth.* Cambridge, Mass.: Harvard University Press, 1950.

Spitzka, Edward Anthony. "A Study of the Brain of the Late Major J. W. Powell." *American Anthropologist* 5 (1903), 585–643.

Spurr, Stephen H. and Burton V. Barnes. *Forest Ecology.* 3rd ed. New York: Wiley, 1980.

Stegner, Wallace. *Beyond the Hundredth Meridian: John Wesley Powell and the Second Opening of the West.* New York: Penguin, 1992. (First published by Houghton Mifflin, 1954.)

———. *Where the Bluebird Sings to the Lemonade Springs.* New York: Penguin, 1992.

Stevens, Larry. *The Colorado River in Grand Canyon: A Comprehensive Guide to Its Natural and Human History.* Flagstaff, Ariz.: Red Lake Books, 1993.

Worster, Donald. "The Legacy of John Wesley Powell." In Hal Rothman, ed. *Reopening the American West.* Tucson: University of Arizona Press, 1998.

———. *A River Running West: The Life of John Wesley Powell.* New York: Oxford University Press, 2001.

———. *Rivers of Empire: Water, Aridity, and the Growth of the American*

West. New York: Oxford University Press, 1985.

———. *An Unsettled Country: Changing Landscapes of the American West.* Albuquerque: University of New Mexico Press, 1994.

INDEX

Note to Index

A page number in bold type indicates the beginning of a chapter or section on the topic. In the body of the index, the name of John Wesley Powell may be abbreviated as "J.W.P."

Advertising, 356

Agricultural schools, 186

Agriculture, 229, 241–42, 353; in the arid lands, 142–43, 145–46, 229–30, 239–40, 256–57; benefits of irrigation to, 3, 166–67, 229, 236; cooperative farms proposed, 143, 145, 191, 310n.11, 310, 312; mining and, 272–73, 298; rainfall and, 157–58

Alcove Creek, 45

Alkaline lakes and valleys, 230

Alluvial cone, 170n.18, 170

American Anthropological Association, 18

American Indians. *See* Indians

Ancestral worship, 126

"Ancient Province of Tusayan, The," 95, **107**

Anthropological Association, American, 18

Anthropological Society of Washington, 325, 340

Anthropological studies. *See* Culture; Ethnographic studies

Apache Indians, 122

Arid lands: agriculture in, 142–43, 145–46, 229–30, 239–40, 256–57; classification of, 150n.1, 191, 207, 278; deserts, 117, 178–79, 222, 230, 295; fertility of when irrigated, 150, 256, 257, 258; geographic extent of, 166, 183, 279–80; high desert and plateaus, 114, 295; irrigation required for agriculture in, 300n.2, 300–301; surveys required to classify types of, 2, 211–12, 309–10.